August Wilhelm von Hofmann

Introduction to Modern Chemistry

Experimental and Theoretic

August Wilhelm von Hofmann

Introduction to Modern Chemistry
Experimental and Theoretic

ISBN/EAN: 9783337038946

Printed in Europe, USA, Canada, Australia, Japan

Cover: Foto ©berggeist007 / pixelio.de

More available books at **www.hansebooks.com**

INTRODUCTION

TO

MODERN CHEMISTRY

EXPERIMENTAL AND THEORETIC.

EMBODYING

TWELVE LECTURES DELIVERED IN THE ROYAL COLLEGE OF CHEMISTRY, LONDON,

BY

A. W. HOFMANN, LL.D., F.R.S., V.P.C.S.

PROFESSOR OF CHEMISTRY IN THE ROYAL SCHOOL OF MINES; ASSAYER TO THE ROYAL MINT; AND EXAMINER IN CHEMISTRY TO THE DEPARTMENT OF SCIENCE AND ART.

LONDON:
WALTON AND MABERLY,
GOWER STREET, AND IVY LANE, PATERNOSTER ROW.
1866

TO

SIR JAMES CLARK, BART., M.D., F.R.S.

PHYSICIAN TO HER MAJESTY THE QUEEN.

MY DEAR SIR JAMES,

Two motives, either of which would be adequate, inspire me with the wish to dedicate this work to you.

In the first place, the lectures it embodies were delivered in an Institution of which, under the auspices of the illustrious Prince Consort, you were one of the principal founders,—and which owed its passage through the perils of infancy mainly to your unremitting and strenuous support.

In the second place, appearing, as this book does, at a moment when duty calls me away to labour in another sphere, it affords a particularly fitting occasion for the acknowledgment of my deep debt of gratitude to one whose sympathy has so kindly and constantly sustained me in endeavouring to promote, in the country of my adoption, the great cause of chemical education.

Not for any intrinsic merit of its own, but as one effort more in that noble cause, you will accept, I am persuaded, my dear Sir James, the dedication of this little work in token of my sincere regard and unalterable friendship.

A. W. HOFMANN.

London, March 25, 1865.

PREFACE.

THE following pages contain, in their latest form, with some additional developments, the introductory portion of the chemical course which the Author has annually delivered, during the last fifteen years, in the theatre of the Royal College of Chemistry.

The juncture of chemical history at which this little book is published, and the peculiar educational necessities which that juncture implies, are reflected as well in the substance as in the method of the work; the purport of which will, therefore, be best made known by a brief reference to the present posture of chemical affairs.

No chemist will need to be reminded that, during the last quarter of a century, the science of chemistry has undergone a profound transformation; attended, during its accomplishment, by struggles so convulsive, as to represent what, in political parlance, would be appropriately termed a Revolution.

Amidst continual accessions of fact, so rapid, so voluminous, and so heterogeneous, as almost to exceed the grasp of any single mind, chemical science has been in travail, so to speak, with new laws and principles of co-ordination, engendered, perhaps, partly by the sheer force of their own deeply-felt necessity, but partly also, and mainly due, to the powerful initiative impulsion of a few philosophical master-minds.

Based on the concurrent examination of the volumetric and

ponderal combining-ratios of certain typical elements, and on the recognition, in their standard combinations, of a few well-marked structural types, these principles have introduced into the domain of chemistry the pregnant idea of *Classification*—the conception of a series of natural *Groups*, resembling the genera of the biological sciences, and culminating in the establishment of an orderly *System*, where before there had seemed to be but a chaos of disconnected facts.

Under the influence of these and certain other cognate ideas, new views have arisen as to the constitution and chemical properties of matter; a reformed chemical notation has thence of necessity ensued; and structural relations, previously unsuspected, have disclosed identity of parentage in compounds till then deemed utterly diverse.

It appears to be wisely ordered, in scientific as in social affairs, that the innovating spirit which belongs to Youth has its check and counterpoise in the conservative tendency essentially characteristic of Age; so that, in the sharp collision of these rival forces, new principles, in any kind, find a sort of fiery ordeal interposed between their first enunciation and final acceptance; doubtless the appointed test of their soundness and vitality.

Hence the domain of chemical philosophy has, for many years past, rather resembled a tumultuous battle-plain, than a field bestowed by nature for peaceful cultivation by mankind. The new ideas, springing up of necessity one by one, and not always free, at their first conception, from errors and inconsistencies, have been resisted, by the champions of the old chemical dogmas, as a gratuitous revolt against established authority. Controversy has naturally stimulated research, which, in its turn, has produced rapid modifications of theory; so that the aspect of chemistry has been in a state of incessant change. It is, indeed, only within the last few years that the new doctrines have acquired a logical consistency, and a consequent ascendency throughout Europe, auguring at length, for our long-agitated science, a period of comparative calm.

The Author's chemical lot, both as a student and as a teacher, has been cast amidst the storms of this controversial period; in which he has felt it his duty to take part on the side of innovation.

During the lengthened period which he has passed in tuition at the College of Chemistry, the task imposed on him has been, to hold the balance fairly between the old views and the new; between authorized conceptions, evidently on the wane, and novel generalizations still awaiting final proof; so that his teachings may have sometimes almost seemed to resemble those dissolving scenes, which, at a certain moment, present two landscapes, one in the act of melting away, while the other is unfolding itself to view.

It will be readily understood that a printed record of such discourses, though frequently and earnestly solicited by pupils, could scarcely have possessed any permanent value, had it been produced amidst the doubts, and reflected the half-truths, of a period so eminently transitional. It will also be felt that its publication has become more opportune, now that a consistent body of doctrine, novel, yet based on irrefragable fact, can be put forth as undoubtedly permanent.

It will be apparent from the above remarks that this work is of an essentially general and introductory character, designed to elucidate the leading principles of chemistry, and by no means presented as an encyclopædic compendium of its facts. So far, indeed, from seeking to multiply details, it has been the author's chief care to avoid them, and to enter upon descriptions of phenomena only in subordination to his main design. This will, perhaps, be most readily gathered from the brief résumé of these lectures, with which the last of the series concludes.

Moreover, as the Author's aim is essentially educational, so he has made it his care to avoid a too systematic treatment of his subjects; preferring the experimental, illustrative method, so peculiarly adapted for the lecture-room. Thus, at the very outset, he plunges in *medias res*; starting abruptly with an experiment; and leaving the principles therein involved to unfold themselves naturally from the explanation of the phenomena displayed. Throughout the course he has adhered to this method; proceeding constantly from the concrete to the general; and extracting from a limited range of facts the largest amount of theoretic and general information which they can readily be made to yield.

In the fulfilment of this programme he has had of necessity to break with the classical traditions of chemical teaching. The elements, for example, are here studied in a new order, not gratuitously adopted, but determined by the Author's view of their fitness to lead up to the knowledge of general laws, in just and logical succession; and so also, while endeavouring to illustrate, incidentally, the leading topics of experimental chemistry, he has been mainly guided, in his selection of experiments, by their fitness for the elucidation of theoretical views.

It is rather in accordance with these exigences of his plan, than with reference to the relative importance of the subjects treated, that space has been meted out to these; some topics, in themselves of great moment, being dismissed with but a passing notice; while others, intrinsically less interesting, are elaborately discussed, on account of their bearing on questions of principle.

The execution of such a project as this is greatly facilitated by the Lecturer's happy prerogatives. The mere limits of time to which he is bound, preclude, in any case, his attempting the exhaustive treatment of his themes. At the lecture-table he is only expected to display a few salient facts, in a striking and attractive form, and to deduce therefrom a few guiding principles, so as to assist his auditors in acquiring for themselves the details of the science. Any attempt on the lecturer's part to make his brief discourses encyclopædic must, of necessity, fail; nay, lectures are probably by so much the better fitted for their purpose, by how much they are freer from unnecessary detail, and more thoroughly emancipated from the trammels of systematic routine.

The Author touches on these points in order that more may not be expected from this book than its very design permits it to supply. Its text, indeed, as originally written, was an almost verbatim reproduction of his actual language at the lecture-table; the experiments described are those which were really shown before the class; and the woodcuts faithfully represent the apparatus employed in their performance.*

Into the original framework thus provided, there has been introduced, in preparing these discourses for the press, such additional matter as appeared desirable for the more complete elucidation of the great laws set forth.

In the performance of this task, which came upon him amidst the almost overwhelming pressure involved in the simultaneous completion of old engagements and preparation for new ones, the Author gladly availed himself of the kindly proffered collaboration of his esteemed friend, Mr. F. O. Ward; whose well-known powers of lucid composition, and habits of philosophical thought, will be traced in every chapter of this work. Attracted to the new chemical doctrines by their own intrinsic truth and beauty, Mr. F. O. Ward has willingly devoted himself, for months past, to the task of assisting in their exposition; and in

* The Author cannot refrain from acknowledging the great pains bestowed by the artist, Mr. Julius Jury, in delineating these illustrations; and by the engraver, Mr. Rexworthy De Wilde, in executing them. His thanks are also due to Messrs. McLeod and Gilman, for their care and skill in making the photographs from which many of the drawings were copied.

the course of these labours, as was to be expected, he has originated many valuable conceptions for their clearer elucidation and development. One, indeed, of his friend's indications the Author feels bound to mention here, as constituting a distinct and valuable contribution to the new chemical edifice. He alludes to the Quantivalential Equilibrium of the Nitroxygen series, as demonstrated by Mr. Ward, and displayed in the striking symmetrical diagram introduced by him at p. 183.

In conclusion, the Author desires to express his heartfelt thanks to Mr. F. O. Ward for his invaluable collaboration.

Royal College of Chemistry, London.
March 24, 1865.

CONTENTS.

LECTURE I.

LECTURE II.

LECTURE III.

LECTURE IV.

LECTURE V.

LECTURE VI.

LECTURE VII.

LECTURE VIII.

LECTURE IX.

LECTURE X.

LECTURE XI.

LECTURE XII.

INTRODUCTION TO MODERN CHEMISTRY,

EXPERIMENTAL AND THEORETIC.

LECTURE I.

Water—its decomposition by the alkali-metals—the resulting gas, hydrogen—its principal characters—its volume-weight—other sources of hydrogen—muriatic acid and ammonia—dissolved in water—disengaged as gases—desiccation of the gases—their distinctive properties—development of hydrogen from muriatic acid and ammonia by the alkali-metals—processes and apparatus—further sources of hydrogen—decomposition of muriatic acid, water, and ammonia, and disengagement of hydrogen therefrom by electric agency.

It is a well-known fact that Water can be brought into contact with many of the metals, without undergoing any sensible change. Gold and silver do not produce the slightest effect upon it; and even copper, iron, zinc, and tin, can be immersed for a considerable time in water at the ordinary temperature of the atmosphere, without causing any change in the condition of the fluid. This does not, however, hold true of all metals.

Many metals produce a very decided action on water. Amongst these we may, for our present purpose, cite two in particular, viz., Potassium and Sodium, the one prepared from wood ashes, the other from common salt, by processes to be explained hereafter. A ball of potassium, thrown upon water (Fig. 1), ignites, and glides with a hissing noise along the surface; emitting a bright violet light

Fig. 1.

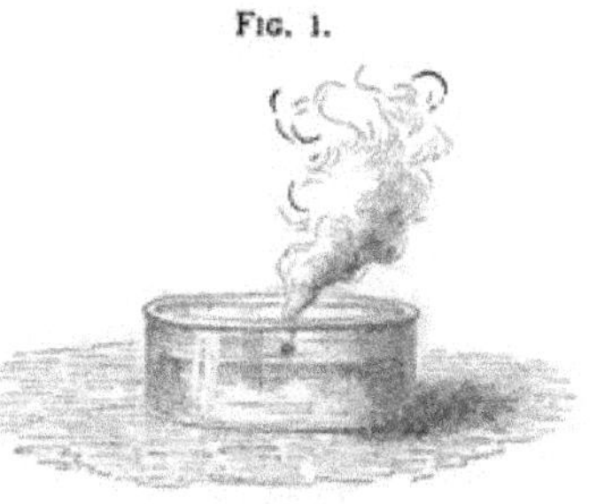

evolving white fumes, and presently disappearing with a slight report. A piece of sodium acts in a similar manner, though the phenomenon is less striking. The floating ball does not ignite unless its motion be arrested; as, for instance, by spreading a sheet of blotting-paper over the water, when the metal burns with a bright yellow flame. In both cases the water acquires a peculiar taste, called alkaline, and the faculty of modifying vegetal colours not affected by pure water. A strip of yellow turmeric paper, if dipped into water that has been exposed to the action of potassium or sodium, turns brown, whilst red litmus paper is changed to blue.

What becomes of the potassium and sodium which seem thus to be consumed by the mere touch of water, and which in reality cease to exist as metals?

What, again, is the nature of that concomitant change in the water itself, to which it owes its acquisition of a new and peculiar flavour, and the power of transforming vegetal colours on which it previously exerted no influence?

These questions, and the class of modifications in the properties of matter to which they refer, belong to the science called Chemistry, a term of obscure origin, traced by some to Χημία, one of the ancient names of Egypt, where the study of these mysterious transformations is supposed to have taken its rise. For the investigation of this class of phenomena, which is the purpose of these Lectures, the striking facts we have just witnessed, and the questions which they suggest, furnish an appropriate starting point; and I will therefore ask you to accompany me in subjecting them to a close and rigorous scrutiny.

FIG. 2.

For this purpose, let a cylindrical glass vessel, closed at

one end, be filled with water, and its open mouth covered with a flat glass plate; let it then be plunged mouth downwards (Fig. 2) into a basin of water, and fixed in such manner that its orifice may be beneath the surface of the water, without touching the bottom of the vessel. The water will, of course, be sustained by atmospheric pressure within the cylinder, so as completely to fill it. Now let a ball of sodium be thrown upon the water (potassium may be employed, but is less appropriate on account of the greater violence of its action), and conveyed, by means of a sort of spoon with a wire-gauze bowl, beneath the mouth of the inverted cylinder (Fig. 3). Colourless

Fig. 3.

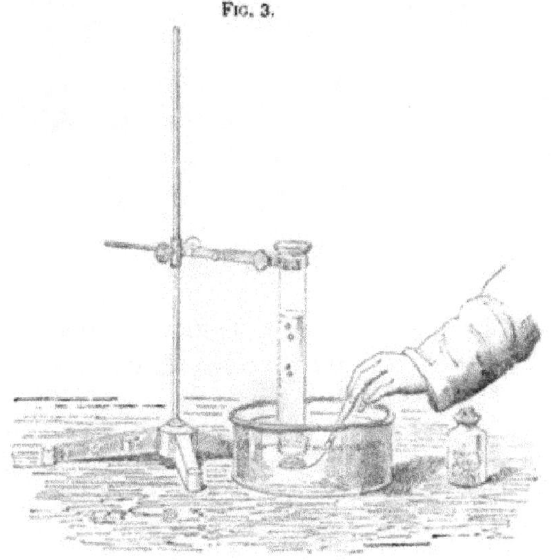

gas bubbles immediately appear, and rise in the cylinder, displacing the water. By repeating this operation three or four times the cylinder may be completely filled with gas. This done, let the glass plate be replaced under the orifice (Fig. 4), and the cylinder be raised out of the water and turned mouth upwards.

The gas which has been thus separated from the water, is called Hydrogen. It is colourless, transparent, and has neither taste nor smell; in these respects resembling common atmospheric air, from which, however, it differs in many particulars.

Thus, for example, if brought into contact with a lighted candle (Fig. 5), hydrogen inflames and burns with a pale lambent flame, quietly descending into the vessel.

Fig. 4.

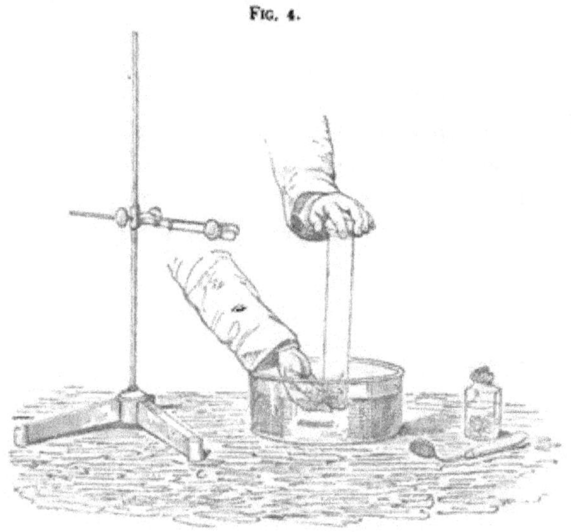

To secure this result, however, the glass plate must not be removed from the mouth of the cylinder till this has been brought

Fig. 5.

close to the burning candle; for if the cylinder be left open only a few seconds, every trace of the inflammable gas will escape, and only common air will remain in its place. If, on the other hand, the glass plate be removed from the cylinder charged with hydrogen, without turning the mouth upwards, the result will be different. In this position twenty minutes or more may elapse before the inflammable gas has dispersed; a fact easily ascertainable by applying the simple test of the lighted candle. Instead of suffering the hydrogen thus to escape into the air, it may be collected in a cylinder held mouth downward over the ascending stream. In this way a jar full of hydrogen may, so to speak, be poured or decanted *upward*, into an inverted cylinder (Fig. 6). We infer from the facility with which the gas escapes when an upward vent is afforded for its egress, that a given bulk of hydrogen is lighter than an equal bulk of common air. Accurate experiments have proved that the weights of equal volumes of hydrogen and of common air are in the proportion of 1 to 14·4; in other words, that the air is 14·4, or nearly 14½ times heavier than hydrogen.

Fig. 6.

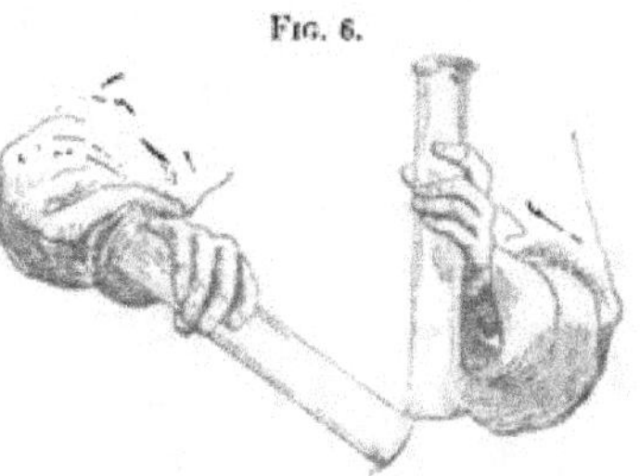

The weighing and measuring of aëriform bodies involves certain precautions which we shall have to consider more fully hereafter; for our present purpose we need only remember that the bulk of an aëriform body depends essentially on temperature and pressure, and that, therefore, equality of these conditions is indispensable to any just comparison of different gases in respect of their *relative volume-weight*, or, as it is commonly termed, their specific gravity.

Of all gases hitherto discovered, hydrogen is the lightest; it has, therefore, been deemed appropriate to take the weight of a given bulk of hydrogen as the unit, in terms of which the weight of an equal bulk of any other gas may be expressed. When, therefore, we say that the specific gravity of the air is 14·4, we mean that a given bulk of it, at given temperature and pressure, weighs 14·4 times as much as an equal bulk of hydrogen under like physical conditions. Hence the densities of gases are necessarily and obviously proportionate to their specific gravities; so that calling the density of hydrogen 1, the density of atmospheric air, like its specific gravity, is 14·4.

Water, however, is not the only substance from which hydrogen can be eliminated by means of potassium or sodium. Two fluids, Muriatic acid and Ammonia, both of which have been known for centuries past, and are at present extensively employed in the arts, yield hydrogen when subjected to the action of the metals above named.

Pure Muriatic acid is a gas, like common air, and like hydrogen. The liquid muriatic acid of commerce is but an aqueous solution of this gas, and gives it forth when heated. This operation is conveniently performed in a glass flask, the mouth of which is closed with a doubly perforated cork; one of the perforations carries a funnel-tube, the lower end of which dips into the liquor, while its upper end widens out to a cuplike form, convenient for charging the vessel. Through the other perforation is fitted a tube bent at right angles (elbow-tube), to give vent to the gas evolved.

The gas issuing from this tube is, however, charged with aqueous vapour, from which it must be freed by contact with a desiccating agent; that is, with a substance more greedy of moisture than itself. Such a substance is that known in commerce as oil of vitriol, termed by chemists sulphuric acid. This acid may be conveniently employed for the purpose of desiccating gases. For this purpose a glass bottle, closed by a doubly perforated cork, is provided. One of the perforations carries an elbow-tube, reaching to the bottom of the bottle; the other perforation is fitted with a delivery-tube. This bottle being filled with pumice-stone saturated with sulphuric acid, and its elbow-tube being attached by a piece of caoutchouc tubing, called a connector, to the elbow-tube of the flask, the gas issuing therefrom is conveyed to the bottom of the acid-soaked pumice-column, through the interstices of which it passes, impinging as it rises against widely-spread surfaces of acid, which greedily absorb its moisture. It thus reaches the upper part of the bottle quite dry, and in that condition escapes from the delivery-tube.

Fig. 7.

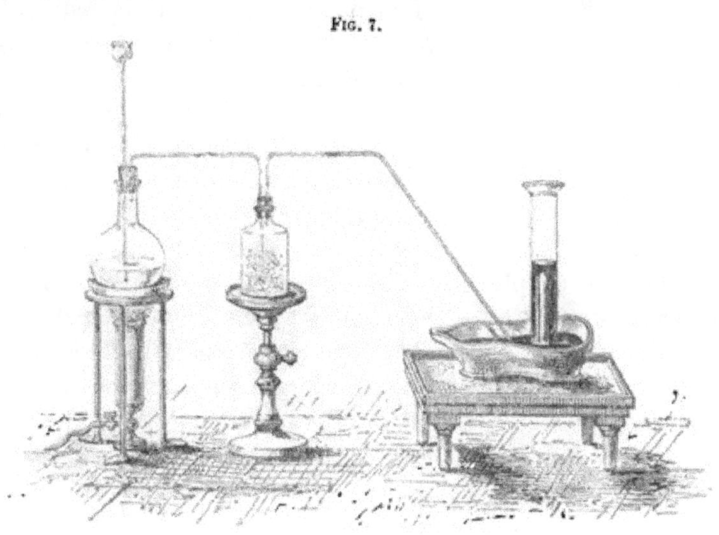

Muriatic acid vapours, thus treated, lose every particle of water, and issue forth a colourless, transparent, and perfectly pure gas. This may be collected in a cylindrical vessel filled with mercury, inverted over a mercurial trough (Fig. 7).

Fig. 8.

Muriatic acid gas is easily distinguished from hydrogen, as well as from common air. It is not inflammable, and, when exposed to a moist atmosphere, it gives rise to the formation of white clouds (Fig. 8). If a cylinder filled with muriatic acid gas be opened under water, the liquid will rush into the gas, as into a vacuum, filling the vessel completely (Fig. 9). The muriatic acid gas is dissolved in the water, and liquid muriatic acid, from which it was in the first instance expelled, is reproduced. Muriatic acid, as well in the form of gas as in a state of solution, affects certain vegetal colours; changing litmus, for instance, from blue to red, as is shown by dipping into it a litmus-stained strip of paper.

Fig. 9.

In order to prove that hydrogen can be procured from muriatic acid by the agency of an alkali-metal, it is only necessary to remove the delivery-tube from the muriatic acid gas apparatus, and to fix in its place, by means of a perforated cork, a tube of difficultly-fusible glass, blown at the middle into a bulb, and containing within the bulb a piece of potassium. These dispositions being made, the muriatic acid gas generated soon reaches the metal, which forthwith becomes covered with a white incrustation, and if the bulb be now very gently heated by a

Fig. 10.

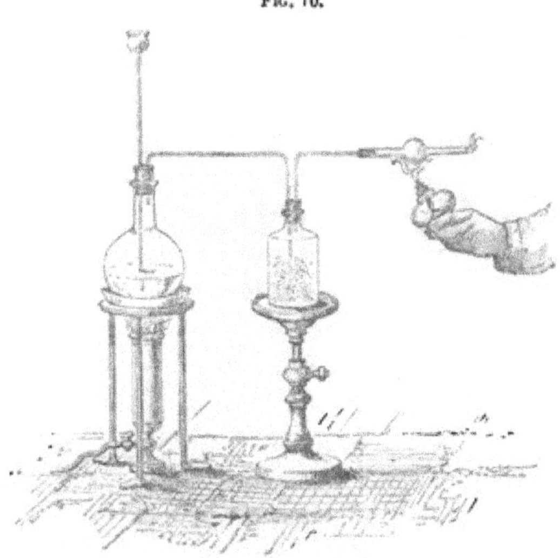

spirit lamp, the potassium fuses, at the same time taking fire, and burning with a violet light. During this action hydrogen is evolved, and may be ignited at the orifice of the tube (Fig. 10).

Sodium produces exactly similar results, but at a much higher temperature. The requisite temperature may, however, be greatly reduced by substituting a solution of sodium in mercury for pure sodium. This solution, known amongst chemists as sodium-amalgam, may be obtained by rubbing the two metals together in a mortar, when they unite with powerful evolution of heat, occasionally rising to actual incandescence. Or mercury may be gently heated in a flask, and sodium added in small fragments, which are dissolved with evolution of heat and light. The minute subdivision of the sodium in the amalgam intensifies the reaction between the metal and the muriatic gas, by multiplying their points of contact, so as to bring about the decomposition of the acid at ordinary atmospheric temperature. If it be desired to perform the experiment on a larger scale, the bulb-tube may be replaced by a flask or bottle filled with sodium amalgam. The gas thus obtained is hydrogen; it

Fig. 11.

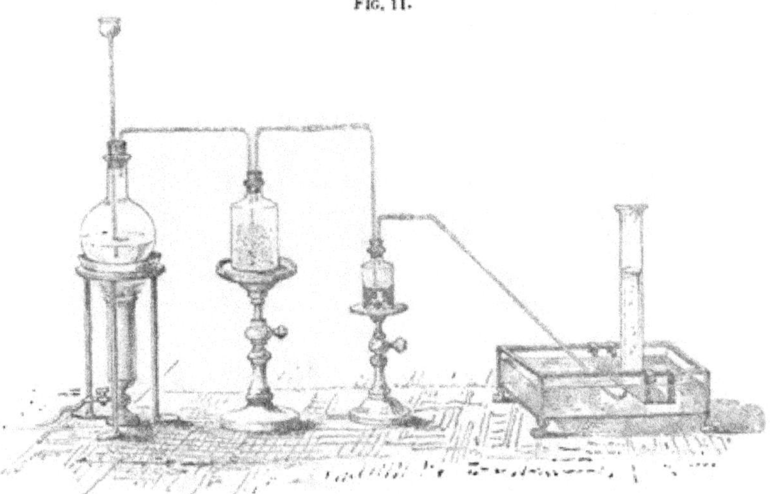

may either be burnt at the orifice of the delivery-tube, or collected over water as before (Fig. 11); and it is easily recognizable by its peculiar properties.

Pure Ammonia, like muriatic acid, is a gaseous body, of which the so-called liquid ammonia is an aqueous solution. This solution, introduced into a flask and moderately heated, evolves the gas abundantly, as its pungent odour attests. The moist ammonia-gas thus generated requires desiccation; for which purpose we pass it through an apparatus analogous to that employed for the drying of muriatic acid gas. For reasons, however, to be explained hereafter, the acid-saturated pumice-column with which we previously filled the drying-bottle is, in this case, replaced by a column of fragments of quick-lime; a substance likewise possessing a strong attraction for water.

The transparent colourless gas that escapes from the lime-bottle is pure dry ammonia-gas, which, like muriatic acid gas,

Fig. 12.

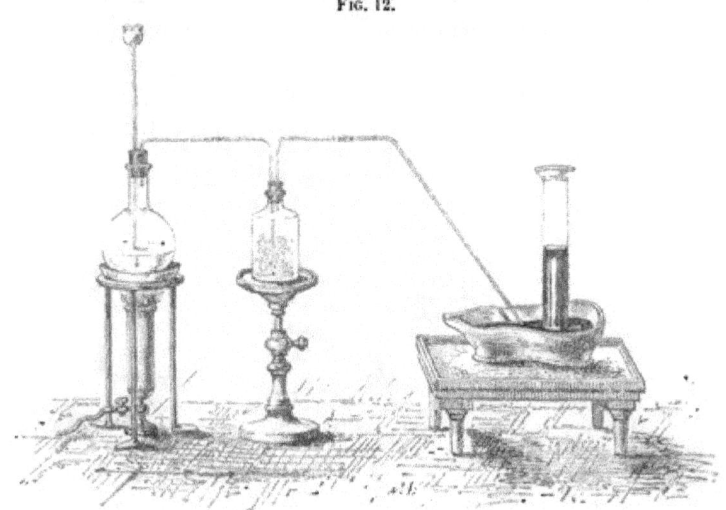

must be collected over mercury (Fig. 12); because water absorbs ammonia-gas (Fig. 13) with even greater avidity than we have seen it display for muriatic acid gas.

From hydrogen, ammonia is readily distinguished by its uninflammability, by its pungent odour, and by its remarkable solubility in water; from air it differs by the two properties last mentioned; from muriatic acid gas, by its odour, by its not reddening vegetal blues, and by its not fuming when brought into contact with atmospheric air. Ammonia-gas is further characterised by the faculty it possesses of restoring to their primitive colour vegetal blues reddened by the action of acids. This property is most readily displayed by dipping into ammonia-gas, or its aqueous solution, a strip of litmus paper, previously reddened by muriatic acid; when the blue tint instantly reappears.

Fig. 13.

The action of potassium or sodium on ammonia may be exhibited by means of the same apparatus as that employed for the analogous demonstration in the case of muriatic acid gas (Fig. 14). For this purpose also potassium is preferable,

Fig. 14.

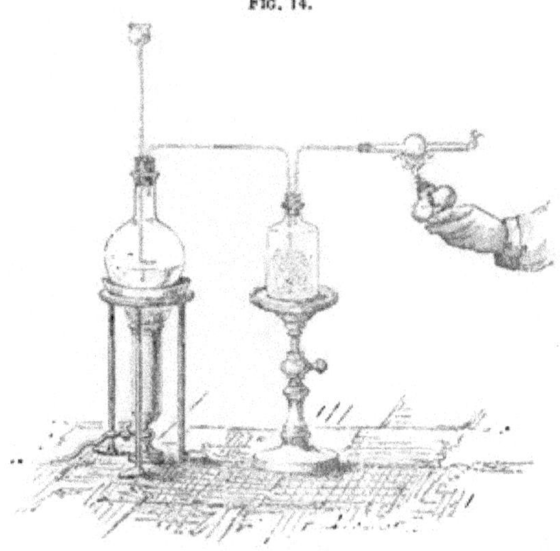

on account of the greater energy with which it acts. So soon as the metal in the glass bulb becomes liquid it is covered with a brownish-green film, and hydrogen begins to escape, as may be proved by lighting it at the mouth of the tube. To collect the hydrogen, we introduce some mercury into a glass cylinder, fill it up with water, and then, closing its orifice with a glass plate, plunge it, mouth downward, into a trough of mercury. The orifice of the delivery-tube being brought beneath the inverted vessel (Fig. 15), the gas rises, first through the mercury, then through the water, into the upper part of the cylinder. During its passage through the water, this solvent frees the hydrogen from such intermixed ammonia as it may have retained. The employment of the mercury is a very necessary precaution. Were the delivery-tube allowed to dip directly into water, the portion of ammonia-gas that had escaped decomposition might, from its powerful attraction for water, cause the liquid to rise in the delivery-tube. The water might thus reach the tube containing the heated metal; whereupon an explosion would inevitably ensue, and shatter the apparatus.

FIG. 15.

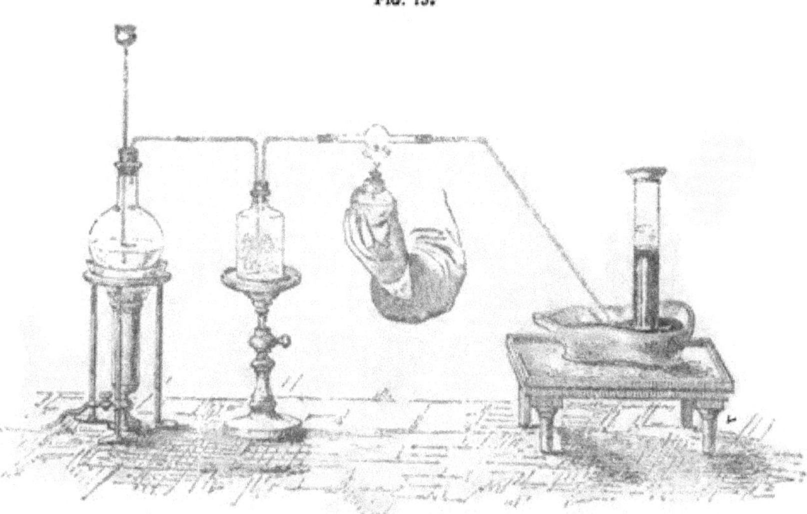

These methods of obtaining hydrogen, by the action of the alkali metals on hydrochloric acid, water, and ammonia, are cited here, not as the best or easiest processes for procuring that gas, but only as illustrative of the nature of those compounds—the subjects of our immediate study. When the chemist's object is merely to procure hydrogen gas in abundance, he resorts to much easier and less expensive processes. Thus, for example, a plate of zinc immersed in liquid muriatic acid, gives rise to a violent evolution of gas, which may be lighted at the mouth of the vessel (Fig. 16), and is thus easily recognized as hydrogen. If the zinc be allowed to act upon the acid in a two-necked bottle, provided with a delivery-tube, the gas may be collected in inverted cylinders over water (Fig. 17). By this process large quantities of hydrogen may be easily and economically prepared. Instead of muriatic acid, lastly, dilute sulphuric acid may very conveniently be used as a source of hydrogen gas. But with these processes we are not at present concerned. For our immediate purpose, it is enough to know that hydrogen may be procured from three different substances—*muriatic acid*, *water*, *ammonia*; and that it may be separated from each of these bodies by the intervention of the same metals, *potassium* and *sodium*.

Fig. 16.

Fig. 17.

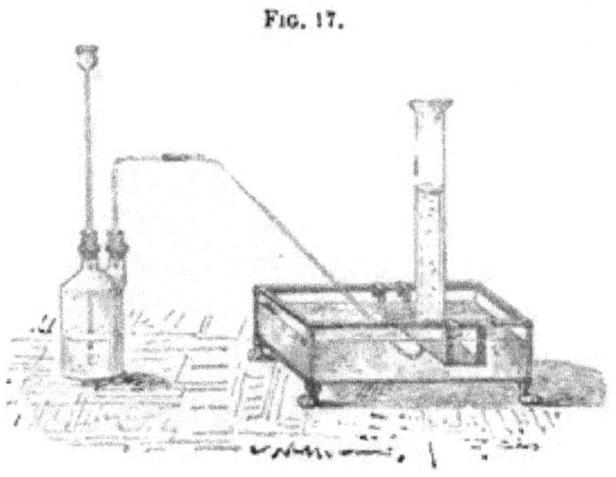

It is, however, proper here to observe that hydrogen may also be obtained from muriatic acid, from water, and from ammonia, by the aid of a peculiar force, called Electricity; which, among many other marvellous properties, possesses that of impressing upon bodies certain modifications of the class which we have already defined as *chemical*.

It does not fall within our present scope to dilate upon the nature of electricity (a name derived from ἤλεκτρον, *amber*, in which resin this force was first generated by friction); still less do our limits permit us to describe the various forms of apparatus by means of which it is developed. The plan of this course debars us from entering on these topics, which belong to the domain of Physics. Indeed, in true logical order, the study of physics should precede that of chemistry; and I might, therefore, fairly assume you to be acquainted with the physical forces and their laws. At all events, it must suffice that we here recall to memory a few particulars which it is essential that we should bear in mind, in order to avoid periphrasis in our future investigations.

The apparatus employed in developing the electric force which we are about to employ is called a *battery;* and the two wires by which the force generated in an electric battery is transmitted and applied, are termed the battery *poles* or *electrodes* (from ἤλεκτρον and ὁδός, meaning roads for electricity). In every battery one pole is termed the *positive*, the other the *negative* electrode; and in the form of battery we shall employ (which consists of zinc and carbon couples), the positive pole proceeds from the carbon, and the negative pole from the zinc end of the battery. As these poles have to be dipped in the fluids operated on, which are often of a corrosive character, it is found advisable to arm them at their extremities with some incorrodible substance, such as platinum or carbon.

At our next meeting we shall proceed to consider the curious and instructive results obtained when the electric force, thus generated and applied, is brought to bear on muriatic acid, water, and ammonia, respectively.

LECTURE II.

Electrical investigation of muriatic acid, water, and ammonia—action of the electric current on muriatic acid—development therefrom of mixed hydrogen and chlorine—separation of chlorine from the mixture—distinctive properties of chlorine—reproduction of muriatic acid, by synthesis, from hydrogen and chlorine—whence its modern name hydrochloric acid—action of the electric current on water—development therefrom of mixed hydrogen and oxygen—separation thereof—distinctive properties of oxygen—liberation of oxygen from water by chlorine—synthetical reproduction of water from hydrogen and oxygen—action of the electric current upon ammonia—evolution therefrom of mixed hydrogen and nitrogen—separation thereof—distinctive properties of nitrogen—its liberation from ammonia by means of chlorine—synthesis of ammonia from its elements not yet achieved—proofs of the composition of ammonia—matter, simple and compound—table of elements.

WHEN a current of electricity is caused to pass through muriatic acid, by immersing therein the poles or electrodes of a battery (Fig. 18), minute gas bubbles are immediately seen rising from

FIG. 18.

their extremities, whilst a peculiar suffocating odour escapes from the fluid. If the experiment be made in a close vessel, these phenomena may be more accurately examined. For this purpose the acid to be operated on is placed in a small cylinder, closed by a cork, through which pass the two electrodes, together with a delivery-tube, so that the gas may be collected, as already explained, in a cylinder inverted over the water-trough (Fig. 19).

Fig. 19.

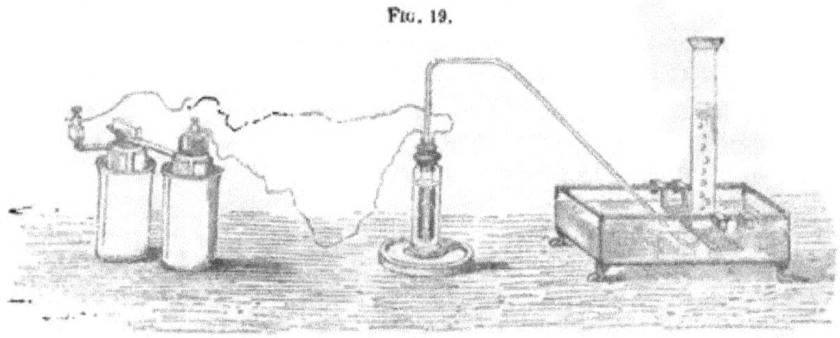

On applying a light to the gas thus obtained, it is found to be inflammable, a fact at once suggestive of the presence of hydrogen. But the peculiar suffocating smell already mentioned does not belong to hydrogen, nor does hydrogen possess the bleaching property which is found to be inherent in the gas escaping from the cylinder. Experimental proof of this bleaching property is readily obtained. For this purpose it suffices to colour the muriatic acid used with a few drops of indigo-solution, which is observed to become rapidly colourless as the gas is evolved; or a strip of litmus paper may be exposed to the current of the gas, which readily bleaches it. It is therefore evident that some gas, as yet unknown to us, is mixed with the escaping hydrogen.

It now, of course, becomes our object to ascertain the nature of the unknown gas, and for this purpose to obtain it alone. Its isolation may be accomplished by means of a V-shaped glass tube, with one closed and one open limb; the former being provided with a platinum-wire fused into the glass, through which it passes to terminate, near the bend of the V, in a slip of platinum-foil. Into this tube muriatic acid of 1·1 spec. grav., coloured with indigo-solution, is introduced, so as to fill the whole length of the sealed, and about half the length of the open, limb. For the purpose of decomposing the acid we connect the negative pole of the battery with the wire of the sealed limb, at the same time inserting the positive pole, through the open mouth of the apparatus, into the liquid (Fig. 20). We now observe that gas is almost exclusively

Fig. 20.

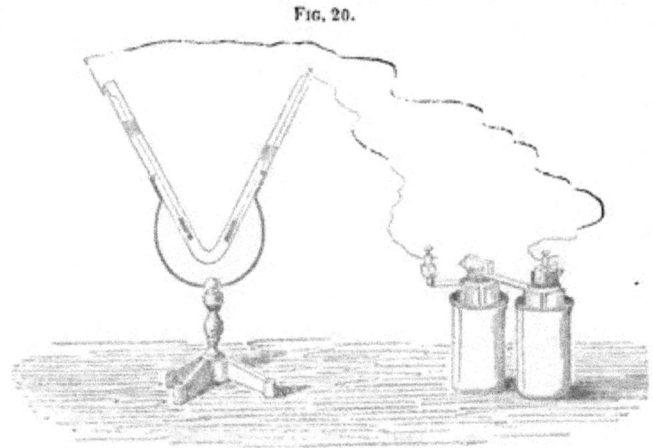

evolved at the negative pole; so slight, indeed, is the disengagement of gas at the positive pole, that it would scarcely attract attention but for its suffocating odour, and its powerful

Fig. 21.

action upon the indigo-coloured solution, which it rapidly bleaches. The gas developed at the negative pole in the sealed limb has no such bleaching powers, but leaves the blue colour of the liquid unchanged. So soon as a sufficient quantity of gas has been collected in the sealed limb (a result usually obtained in eight or ten minutes), the electric current is interrupted, and the gas is transferred to the open limb, previously filled up with water and closed with the thumb (Fig. 21). It is found to be inflammable, and we recognize it without difficulty as hydrogen.

The experiment is now reversed by connecting the positive pole with the sealed, and the negative with the open, limb (Fig. 22);

Fig. 22.

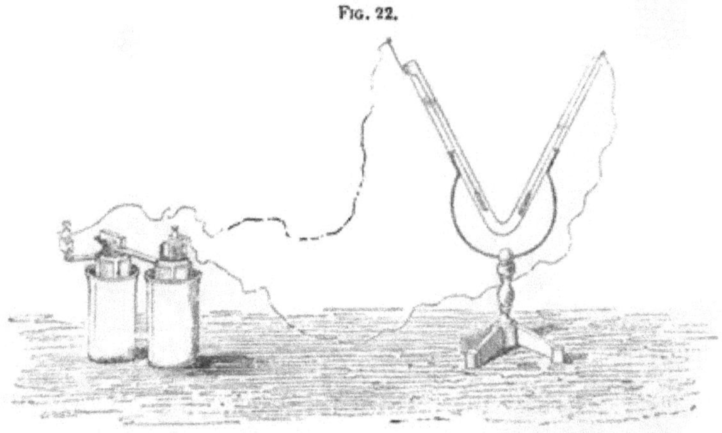

and immediately, as might be expected, hydrogen escapes in abundance from the open mouth of the bent tube, while the liquid becomes decolorized in the closed limb. But this altered disposition of the experiment enables us to acquire some further information regarding the second gas,—that, namely, which is evolved in small quantity, but marked by its powerfully irritating odour, and by its bleaching action upon vegetal colours. In the course of ten or fifteen minutes the decolorized liquid in the sealed limb begins to assume a yellowish-green colour, and the evolution of gas (at the outset scarcely perceptible) becomes gradually more and more copious; so that, in thirty or forty minutes, the greater portion of the tube is filled with a transparent yellowish-green gas. The battery current is now interrupted, and the gas transferred to the open limb for examination. The approach of a light proves it to be uninflammable: the mouth of the tube is no sooner unclosed than its suffocating odour, already mentioned, becomes most plainly perceptible, while its powerful bleaching property is shown by its decolorizing, in the very act of transference to the open limb of the tube, the portion of the acid which had retained its blue colour. This peculiar gas has received the name of Chlorine, derived from the Greek word χλωρός (yellowish-green).

Chlorine may be obtained from muriatic acid by another process, which has the great advantage of evolving it without giving rise to a simultaneous disengagement of hydrogen. Muriatic acid, when heated with powdered black oxide of manganese, in a flask (Fig. 23), yields abundance of chlorine, which

Fig 23

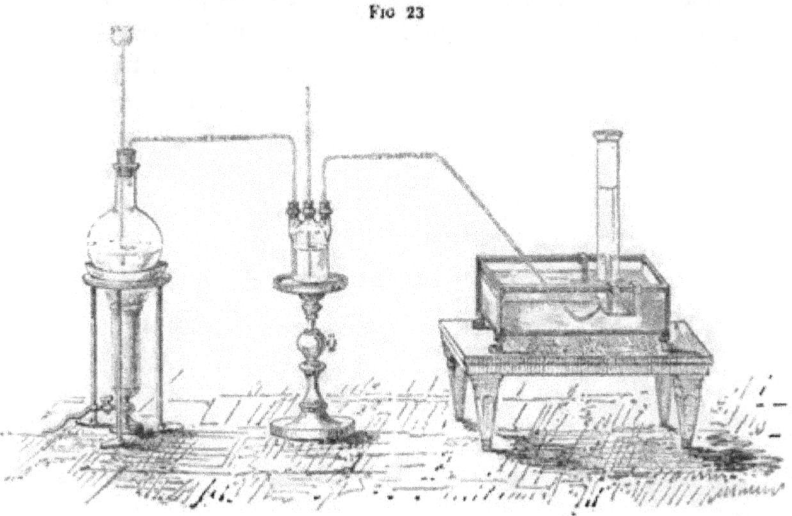

may be collected in inverted gas-jars, over *tepid* water. The manganese process is universally adopted when large quantities of chlorine are required. Hereafter, when we come to study more minutely the nature of this important body, we shall have occasion to dwell on the details of this process: at present it only claims our attention, in passing, as an easy and economical mode of evolving chlorine in abundance, for the purpose of studying its remarkable properties.

Chlorine is soluble in about one-third of its volume of cold water: a property which explains its slow and scanty evolution, in the form of visible bubbles, at the outset of the experiment, and the more rapid disengagement of the gas at a later period, when the liquid is saturated therewith. This solvent power of water for chlorine is, however, greatly diminished by heat; whence the recommendation to collect the gas over *tepid* water. A lighted taper, when introduced into a vessel containing chlorine, burns with a dim flame, depositing a good deal of soot. Of this circumstance advantage may be taken to make manifest the high volume-weight of this gas relatively to that of air and, *à fortiori*, of hydrogen. When a cylinder filled with chlorine is held mouth downward, the colour of the gas rapidly disappears, and its odour is no longer perceptible. If a lighted taper be now introduced into the cylinder, its unimpeded combustion indicates that the chlorine has been displaced by air. Again, if a taper be allowed to burn at the bottom of an air-filled vessel

Fig. 24.

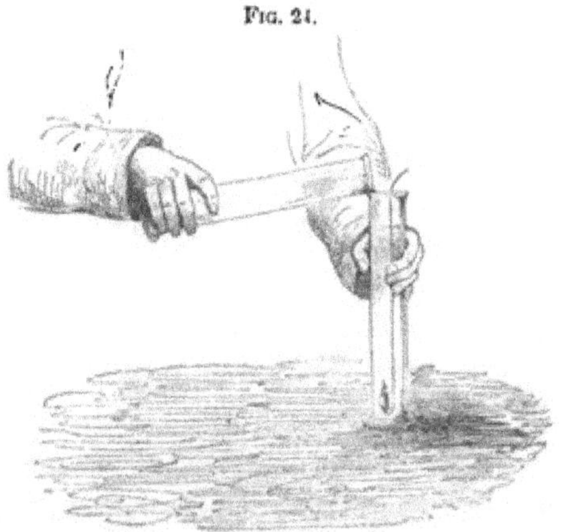

(Fig. 24), and the mouth of a cylinder filled with chlorine be inclined towards it, as if water were being poured, the flickering and smoky flame shows that chlorine is falling on the taper. Thus we learn that chlorine is heavier than air, and consequently very considerably heavier than hydrogen. Exact experiments have proved that, bulk for bulk, chlorine is 35½ times heavier than hydrogen; in other words, that, if the volume-weight of hydrogen be expressed by 1, that of chlorine is represented by 35·5.

From these experiments we have learned that the action of the electric current upon muriatic acid gives rise to the evolution of two essentially different gases; of which, one, hydrogen, familiar to us from our previous experience, is disengaged at the negative pole, while the other, chlorine, our new acquaintance, makes its appearance at the positive pole of the battery. We know, moreover, that each of these gases may be separately evolved from muriatic acid; the hydrogen by sodium, the chlorine by black oxide of manganese; and we are therefore justified in considering hydrogen and chlorine as constituents of muriatic acid.

That muriatic acid contains no other than these two constituents remains to be proved by a further experiment. For this purpose it is necessary to obtain a mixture of the two gases in the proportions in which they combine to form muriatic acid. Such a mixture is most readily obtained by the decomposition of muriatic acid itself into its elements; a decomposition readily effected by the action of the electric current, or, as it is termed, by *electrolysis* (the terminal, *lysis*, being derived from the Greek λύω, I loosen). Let this gaseous mixture be collected over warm water in a glass-stoppered gas-cylinder (Fig. 25), care being taken to allow a considerable quantity of the gas to escape before commencing the collection. Let the cylinder, when full, be closed, and allowed to stand for several hours in diffuse daylight, and ultimately be exposed to the direct influence of the solar beams. Under this treatment the gas will

FIG. 25.

be found to have entirely lost its yellowish tint, and to be no longer capable either of taking fire, or of bleaching vegetal colours. In contact with common air the colourless gas forms white clouds; blue litmus paper exposed to its influence turns red; and if the cylinder containing it be opened under water the gas is rapidly absorbed. These are the characters of muriatic acid; and it is evident, therefore, that the hydrogen and chlorine have, in this experiment, reunited to form once more the substance from which they were originally separated. We arrive, of course, at exactly the same result if we make use of hydrogen and chlorine developed from muriatic acid by other means than electricity; as, for instance, by the action of sodium and of oxide of manganese.

By whichever of these methods the two gases are obtained, let them be collected in two separate cylinders; let these be placed one over the other, the lower one upright, the upper one inverted, so that their orifices may meet; then let the glass plates by which they are closed be withdrawn (Fig. 26), and the

Fig. 26

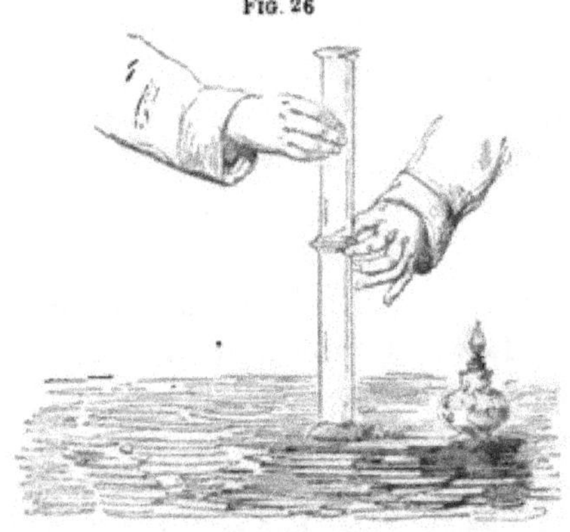

cylinders be shaken, so as to promote the intermingling of the gases; lastly, let the cylinders be separated, so as to admit of a flame being presented to their orifices (Fig. 27), and then it will be seen that the mixed hydrogen and chlorine, at the touch of fire, instantaneously combine. A sort of hissing explosion attends this action, the flame descends into the vessels, and from these dense clouds of muriatic acid escape into the air.

Fig. 27.

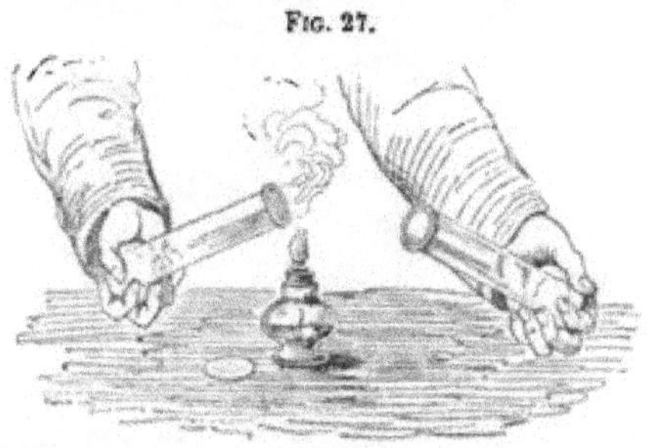

We have thus endeavoured to determine the nature of muriatic acid by two processes; firstly, by decomposing it into its constituents, secondly, by reproducing it from the constituents previously separated. The former process, that of decomposition, is termed the *analytical* method (from the Greek ἀναλύω, I unloose); the second, that of recomposition, the *synthetical* method (from the Greek συντίθημι, I put together). By the analytical method we have found hydrogen and chlorine to be constituents of muriatic acid; by the synthetical method we have proved hydrogen and chlorine to be its *only* constituents, from which circumstance indeed it derives its ordinary chemical appellation, *hydrochloric* acid.

We will now apply these two methods, analysis and synthesis, to the better investigation of the nature of water; of which we as yet only know that, like hydrochloric acid, it evolves hydrogen when submitted to the action of sodium.

The experience acquired in the study of hydrochloric acid points out the steps we have to take for the analytical investigation of water. On immersing the platinum poles of the battery in water, to which a little sulphuric acid has been added for the purpose of increasing its conducting power, we obtain at once, in the copious evolution of gas which ensues, abundant proof of the powerful action exerted by the battery upon the liquid (Fig. 28). If the current be caused to act upon the water in a close vessel, such as, for instance, the small cylinder

FIG. 28.

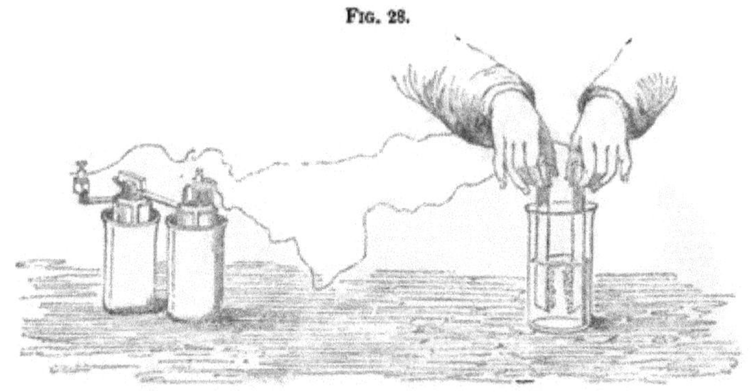

(Fig. 29) fitted up for the analogous experiment on hydrochloric acid, we produce and collect a transparent colourless gas, the inflammability of which suggests the presence in it of hydrogen. But the explosive violence with which the gas burns, the sudden flash of the flame to the very bottom of the cylinder, sufficiently indicate that the hydrogen evolved in this process is mixed with some other gas. For the purpose of

FIG. 29.

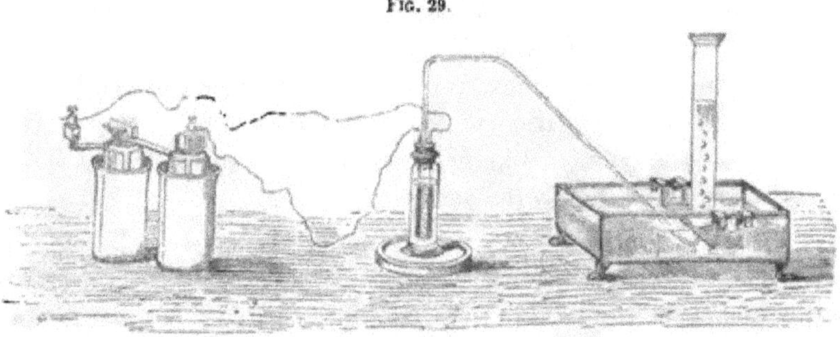

separating the two gases, we resort again to our V-shaped tube, which, having filled with acidulated water, we connect, by the platinum wire at its closed end, with the negative battery pole, at the same time inserting the positive pole in the open limb

(Fig. 30). Streams of gas bubbles appear simultaneously and copiously at both poles, but more profusely at the negative

FIG. 30.

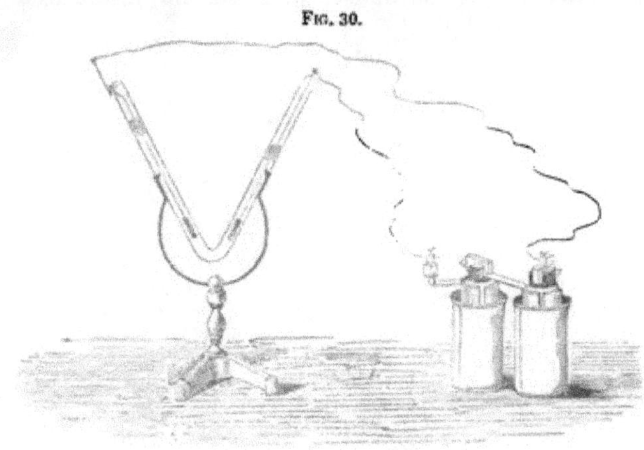

pole. Examination of the gas collected in the sealed limb shows that it consists (exactly as in the similar experiment on hydrochloric acid,) of hydrogen (Fig. 31). The experiment is

FIG. 31.

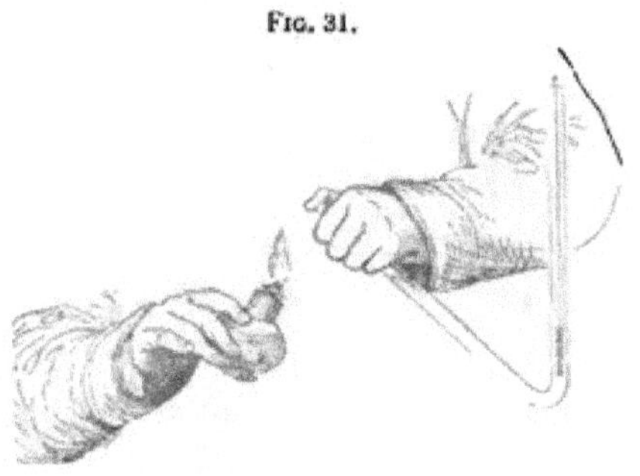

now repeated with the electrodes reversed (Fig. 32) (as before), and the hydrogen disengaged at the negative pole allowed to escape into the air. The gas previously evolved at the positive pole in the open limb, and consequently lost, is now collected in the sealed end of the apparatus.

This gas, like hydrogen, is transparent and colourless; but

it proves on examination to differ essentially from hydrogen. It is not itself inflammable, but it gives intense brilliancy to the combustion of burning bodies, as, for instance, to the

FIG. 32.

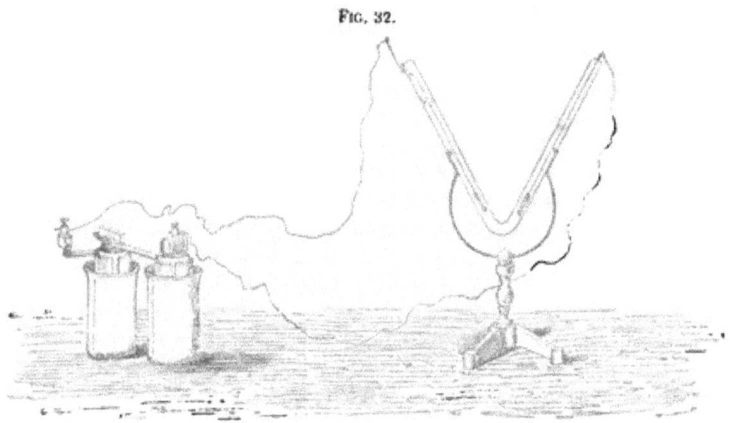

flame of a wax taper immersed in it (Fig. 33). A wooden splinter, retaining but a single ignited spark, when brought into contact with this gas, instantly exhibits a vivid incandescence,

FIG. 33.

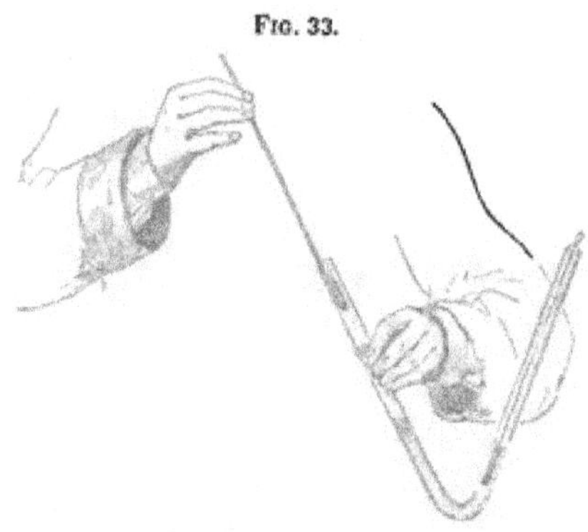

and in the next moment bursts into flame. This gas, which has received the name of Oxygen, is marked, more strongly than either of the gases which we have hitherto examined, by the characteristic properties of common air. It may, indeed, be incidentally mentioned here that this gas is a principal constituent of air.

Oxygen gas is somewhat heavier than atmospheric air. This is readily proved by a simple experiment. Let two cylinders be filled with oxygen, and one of them be placed mouth upward, while the other is suspended mouth downward; both being uncovered. After the lapse of a few minutes, let the lighted taper, or the faintly-glowing wood splinter, be brought successively to the mouth of each jar to test the quality of its contents. The ensuing sudden and vivid combustion will prove the upright cylinder still to contain oxygen, while the absence of these phenomena will show that the oxygen previously contained in the inverted jar has flowed downward through the lighter air, which has ascended to take its place in the cylinder. There is, however, but little difference between the weights of equal volumes of air and oxygen. The relative volume-weight or specific gravity of oxygen, as determined by accurate experiment, is represented by 16, that of common air being 14·4, if the volume-weight of hydrogen be taken as the unitary standard of comparison = 1.

There are numerous methods by which oxygen may be prepared more conveniently and abundantly than by the action of the electric current on water. For the present, however, we must defer the examination of all these processes save one, which is specially suited to our purpose, as it enables us to extract the oxygen from water without simultaneous disengagement of hydrogen. Whilst studying hydrochloric acid, we had occasion to observe how great an attraction exists between hydrogen and chlorine, and how readily these two gases, by their combination, reproduce hydrochloric acid. Mindful of this behaviour, are we justified in presuming chlorine to be capable, under favourable circumstances, of liberating oxygen from water by withdrawing its hydrogen, and combining therewith to form hydrochloric acid? We know that, at the ordinary temperature, this action does not occur, for we have collected chlorine over water. Nevertheless, the decomposition in question takes place readily at high temperatures. This may be shown by means of an arrangement of apparatus, which, though somewhat complex in appearance, is, in principle, simple and easily understood. In the larger flask (Fig. 34), chlorine is evolved from hydrochloric

FIG. 34.

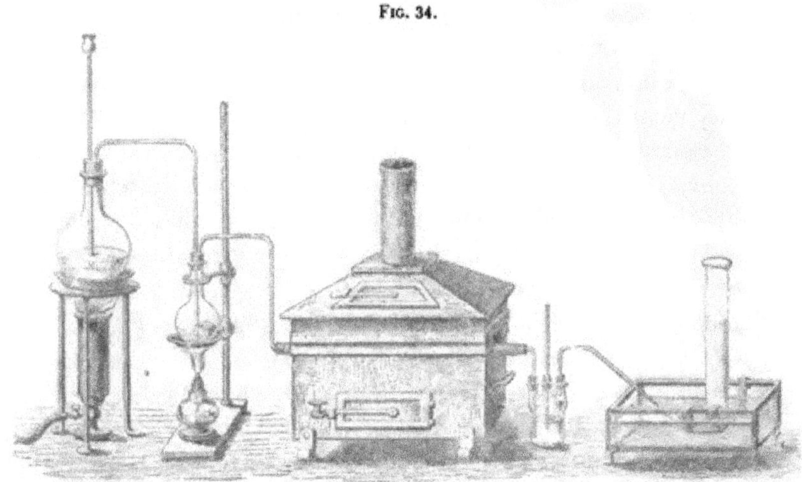

acid by means of black oxide of manganese, and the gas escaping from this flask is caused to bubble through hot water contained in the smaller flask. The chlorine, thus saturated with steam, is then passed through a porcelain tube heated to redness in the furnace. The gas which escapes from the red-hot tube is easily recognized as a mixture of oxygen and hydrochloric acid. To separate these two gases it is only necessary to connect with the furnace-tube a wash-bottle filled with water (or solution of soda), which absorbs and retains the hydrochloric acid, allowing the pure oxygen to pass on through the delivery-tube into the inverted cylinder.

We have thus endeavoured to establish the composition of water by *analysis*; that is to say, having on the one hand simultaneously evolved hydrogen and oxygen from water by means of the electric current, and liberated, on the other hand, the hydrogen by means of sodium, and the oxygen by means of chlorine, we have analytically found hydrogen and oxygen to be constituents of water.

In order to prove, by the converse or *synthetical* method, that hydrogen and oxygen are the *only* constituents of water, we have, as already explained, to reproduce the water from hydrogen and oxygen.

For this purpose we employ a two-necked glass bottle, having a funnel-tube fixed in one of the necks, and a vitriol-charged desiccating-tube in the other; a delivery-tube, drawn out to a fine orifice, terminates the desiccator. Hydrochloric acid is introduced into this bottle, and zinc is added (comp. p. 13); hydrogen is thus liberated, and, losing its moisture in its passage through the desiccator, issues perfectly dry from the terminal jet. Having allowed the gas to escape for some time, we apply a light to the issuing stream, and introduce the burning jet into a bell-jar filled with dry oxygen gas (Fig. 35).

FIG. 35.

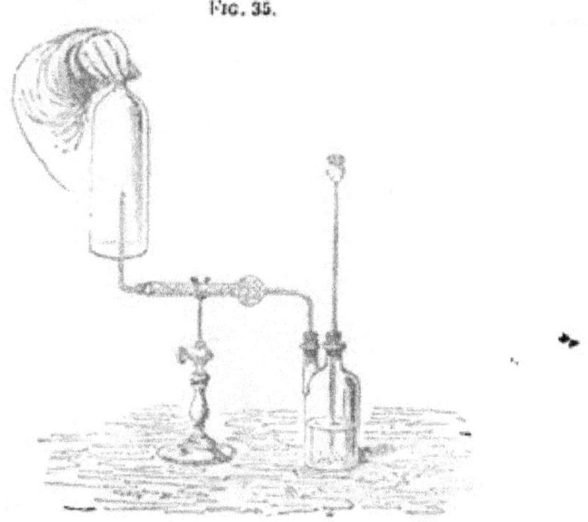

We observe that the walls of the vessel become immediately bedewed with a film of fluid, which gradually collects into droplets, and which we readily recognize as water.

Having thus experimentally investigated the association of

hydrogen with chlorine and oxygen respectively in hydrochloric acid and water, we may now pursue the same inquiry with reference to the constitution of ammonia, the third hydrogen-yielding substance to which our attention has been called.

FIG. 36.

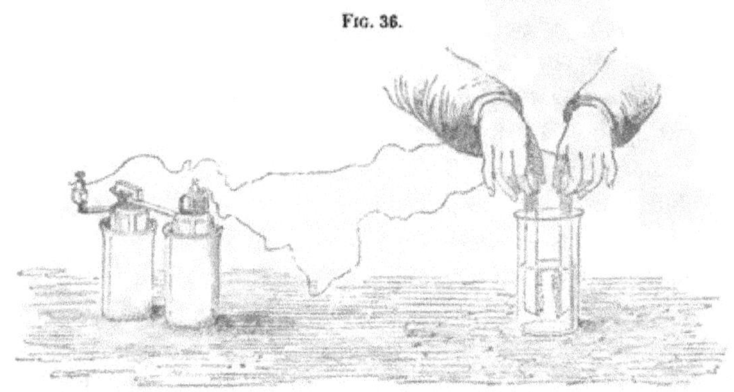

Ammonia may, like hydrochloric acid, be decomposed by electrolysis; and although the action of the electric current is, in this case, somewhat less simple, it is by no means less instructive. The solution of ammonia to be operated on is adapted for the purpose by the addition of a few drops of sul-

FIG. 37.

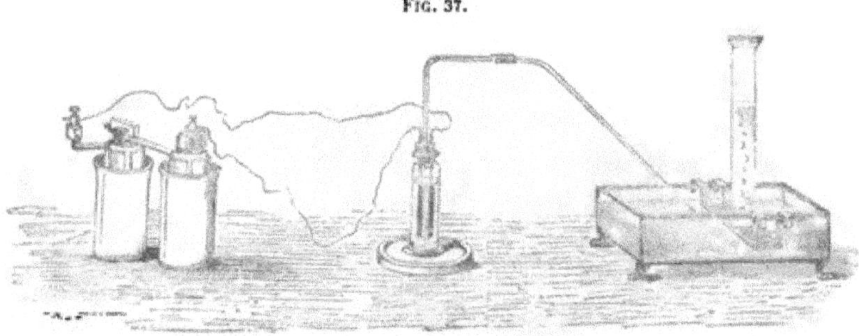

phuric acid to increase its conducting power. The electrodes are then immersed in the liquid, and an effervescence, occasioned by the evolution of gas, is immediately observed to take place

(Fig. 36). On allowing the action to proceed in a close vessel, provided with a delivery-tube (Fig. 37), we collect over water a transparent, colourless, inflammable gas, which we are inclined to consider as hydrogen. But inferring, from the results obtained in the examination of hydrochloric acid and water, the probability that a second gas is present, we have recourse once more to our V-tube, which admits of the gases evolved at the two poles being separately collected. We commence, as

Fig. 38.

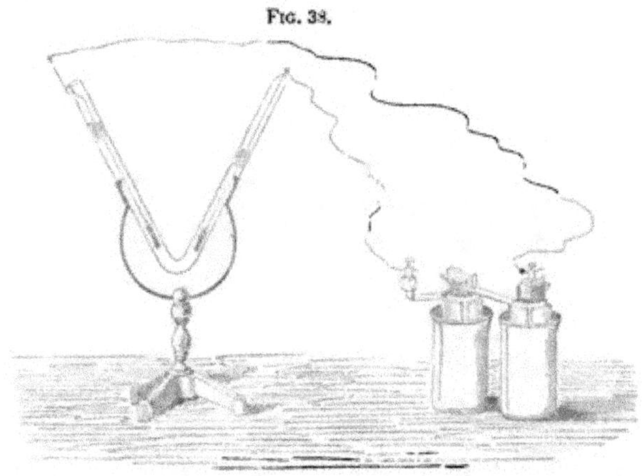

before, by connecting the negative pole with the sealed limb of the apparatus (Fig. 38). The gas, which rapidly collects in

Fig. 39.

the tube, we unhesitatingly recognize as hydrogen (Fig. 39), and we hasten therefore to repeat the experiment with the poles reversed (Fig. 40). We are at once struck with the comparatively small quantity of gas evolved at the positive pole, the process having to be continued for half an hour at least before the gas has been collected in sufficient quantity for examination.

FIG. 40.

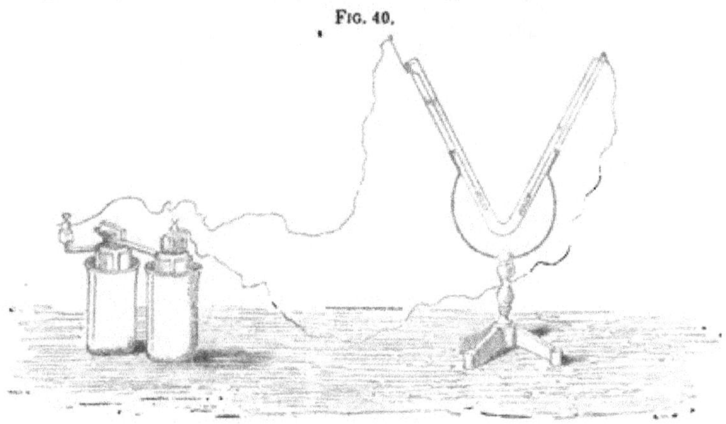

In aspect this gas is transparent and colourless, like hydrogen, from which, however, the approach of the lighted taper proves it to differ essentially in not being inflammable. Nor is it less different from chlorine and oxygen; the absence of colour and odour as clearly distinguishing it from the former gas, as its behaviour towards burning bodies does from the

FIG. 41.

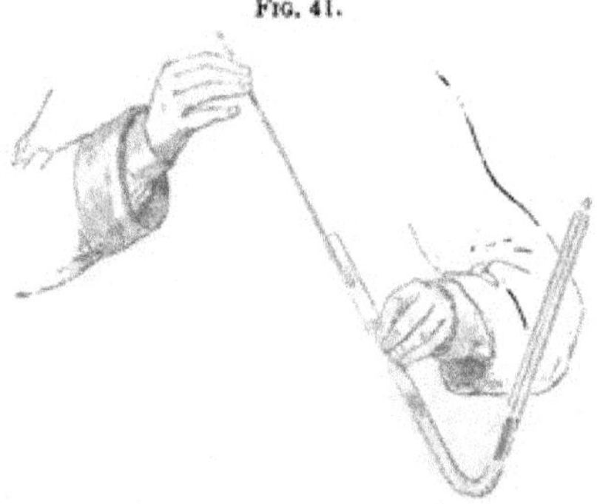

latter. The flame of a taper, when immersed in this gas (Fig. 41), is immediately extinguished. This body, which chemists term Nitrogen, is marked rather by the absence of salient characteristics than by any active properties of its own. Even its volume-weight is, so to speak, of a neutral average character. While hydrogen is very much lighter, chlorine very much heavier, and even oxygen somewhat heavier than air, we find the volume-weight of nitrogen and air to be almost identical. Compared with hydrogen as a standard unit, the volume-weight of nitrogen is 14, that of air being 14·4, as already stated. This will not, however, be matter of surprise hereafter; when, studying in its turn the composition of atmospheric air, we find nitrogen to be its most abundant constituent.

Nitrogen may be liberated from ammonia by a process similar to that which enabled us to separate oxygen from water, namely, by the action of chlorine. Even at common temperatures chlorine combines with the hydrogen of ammonia, and sets free its nitrogen. For this purpose we pass a current of chlorine gas through the strongest ammonia-solution of commerce contained in a large three-necked bottle (Fig. 42). Powerful action

Fig. 42.

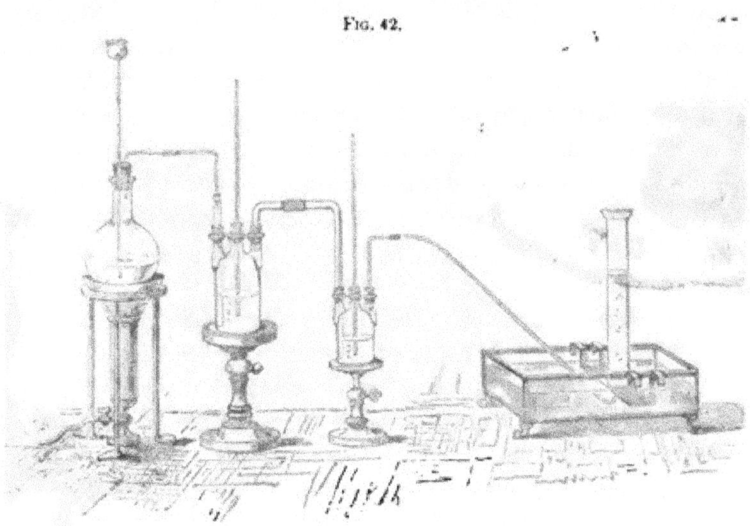

is at once manifested by the formation of white fumes in the upper part of the bottle, while flashes of light traverse the liquid, which is caused to effervesce by the evolution of gas. The white fumes evolved indicate the formation of a solid body, with which we are not at present concerned, further than to mention that it necessitates the use of the wide connecting tubes shown in the figure, as small ones would be speedily choked by the condensation of the fumes. The gas obtained is passed through a wash-bottle, and then collected over water for examination; which soon proves it to be neither hydrogen nor chlorine nor oxygen, but the same gas (nitrogen) which we obtained by the electrolysis of ammonia.

By means of the electric current, we have analytically proved hydrogen and nitrogen to be constituents of ammonia. The presence of these two gases in ammonia has been, moreover, demonstrated; that of hydrogen by the action of sodium, that of nitrogen by the action of chlorine. To proceed in accordance with the plan adopted in the examination of hydrochloric acid and water, we should now by synthesis prove hydrogen and nitrogen to be the *only* constituents of ammonia. Unfortunately, up to this hour no simple process has been discovered whereby ammonia can directly be reproduced from hydrogen and nitrogen. We must, therefore, for the present rest contented with the fact that the united weights of hydrogen and nitrogen extracted from ammonia have been found to correspond exactly with the weight of the ammonia operated on—a result affording irrefragable proof that these two bodies, and no others, enter into the constitution of ammonia.

The study of hydrochloric acid, water, and ammonia, has acquainted us with a series of facts, the importance of which will not be fully apparent to us till we are further advanced in our inquiry. But we have already obtained some partial insight into the wide field of research they open up; and a brief retrospective glance may be appropriate here to prepare us for proceeding with our investigation.

Under the influence of electricity, of heat, and of certain chemical agents, we have seen a small number of well-known substances pass through a series of most remarkable transformations. By appropriate treatment, hydrochloric acid has been found to split up into hydrogen and chlorine, water into hydrogen and oxygen, ammonia, lastly, into hydrogen and nitrogen.

The ingredients, thus dissevered, we have found ourselves able, in the case of hydrochloric acid and water, directly to recombine, so as to call again into existence the compounds we had previously decomposed; and though, in the case of ammonia, direct synthesis proved to be impossible, chemists have ascertained, by means of the balance, the precise weight of each constituent entering into the composition of a weighed quantity of the compound. In this manner, by a twofold demonstration the most cogent that can be conceived, hydrogen and chlorine, hydrogen and oxygen, hydrogen and nitrogen, have been proved to be the true and *only* constituents, respectively, of hydrochloric acid, water, and ammonia.

Having thus succeeded in determining the constituents of hydrochloric acid, water, and ammonia, we are naturally led to inquire into the characters and composition of those constituents themselves.

Is it in our power to resolve hydrogen, chlorine, oxygen, and nitrogen into simpler forms of matter? and, if so, what are the methods of analysis by which this result may be achieved?

To these questions, which have been experimentally propounded to Nature by many of the most illustrious philosophers, as well of the present as of past generations, but one answer has been obtained, viz., that hydrogen, chlorine, oxygen, and nitrogen, are incapable of decomposition by any means as yet at our disposal. They resist the powerful influences of electricity and of heat, even when raised to the highest attainable degrees of intensity; and they issue unchanged from every variety and form of chemical reaction hitherto devised in the hope of resolving them into simpler forms of matter. We are therefore justified in regarding hydrogen, chlorine, oxygen, and nitrogen as *indecomposable*, or *simple bodies*, termed

ELEMENTS, in contradistinction to compound, decomposable bodies, such as hydrochloric acid, water, and ammonia.

How many of these bodies, simple and compound, are there? Of compound bodies, we find a series, numbering many thousands, in the so-called *inorganic kingdom* of nature, which comprises all the diversified mineral constituents of our earth's crust; while another series, far more complex in composition, and almost innumerable in multitude, exists in the two great provinces, vegetal and animal, which together make up Nature's *organic realm*. Yet all this boundless variety of matter springs from only 61 primary bodies, or *elements*, diversely combined. These appear to be the constituents of the celestial bodies, as well as of our earth. The meteorites which at intervals reach our planet from the sky are built up of no other ingredients; and, as a bold but well-founded induction has, of late years, justified the belief that several of them are also constituents of the sun and other fixed stars, we are entitled to regard it as possible that many, if not all, of the remainder may also exist in those bodies.

These 61 elements are arranged alphabetically in the following table, in which, it will be observed, they are classified in three groups, by the use of types of three degrees of prominence. The first group, distinguished by the largest type, comprises, besides the four elements with which we have become acquainted in these introductory sketches, some 14 more, which may rank with them as the 18 most widely diffused bodies found at the surface of our globe. This class includes the main constituents of the Ocean (oxygen and hydrogen), of the Atmosphere (oxygen and nitrogen), and of the Earth's crust (oxygen in combination with silicon, carbon, and the metallic constituents of the earths and alkalis). With these are also classified others which, like bromine and iodine for example, though much less abundant, are equally pervasive. Next in typographical prominence are ranged some 23 elements, comparatively rare, but all more or less useful in the arts, and many of them, such as copper, tin, zinc, &c., familiar to us in every-day life. A third series, distinguised by the smallest type, comprises about 20 elements, which may be termed Nature's chemical rarities—bodies occurring so seldom, and in quantities so minute, as to baffle our endeavours to trace the service which they may render, either in the household of nature or in the arts of life. In this group, three other supposed elementary bodies (Erbium, Terbium, and Norium), would find their appropriate places, had not recent researches shown their existence to be so problematic, as to render them quite inadmissible in a list of established elements.

ALPHABETICAL TABLE OF THE ELEMENTS.

ALUMINIUM.	IODINE.	RUBIDIUM.
ANTIMONY.	IRIDIUM.	RUTHENIUM.
ARSENIC.	IRON.	SELENIUM.
BARIUM.	LANTHANIUM.	SILICON.
BERYLLIUM.	LEAD.	SILVER.
BISMUTH.	LITHIUM.	SODIUM.
BORON.	MAGNESIUM.	STRONTIUM.
BROMINE.	MANGANESE.	SULPHUR.
CADMIUM.	MERCURY.	TANTALUM.
CAESIUM.	MOLYBDENUM.	TELLURIUM.
CALCIUM.	NICKEL.	THALLIUM.
CARBON.	NIOBIUM.	THORIUM.
CERIUM.	NITROGEN.	TIN.
CHLORINE.	OSMIUM.	TITANIUM.
CHROMIUM.	OXYGEN.	TUNGSTEN.
COBALT.	PALLADIUM.	URANIUM.
COPPER.	PHOSPHORUS.	VANADIUM.
DIDYMIUM.	PLATINUM.	YTTRIUM.
FLUORINE.	POTASSIUM.	ZINC.
GOLD.	RHODIUM.	ZIRCONIUM.
HYDROGEN.		

The typography of this table makes it apparent at a glance that only about one-third of the elements are entitled to rank as of primary, and a somewhat similar number as of secondary importance. It is to these two classes of elements, of course, that our attention will be principally directed. On the "chemical rarities" we shall bestow but a cursory glance.

From the manner in which we have arrived at the idea of element, it is obvious that this term must be used with certain restrictions. The bodies enumerated in the above table we consider to be elements, because, with the knowledge at our disposal we have not at present the means of decomposing them. Possibly the progress of science may reveal these means to a future generation, so that bodies which for us are elements may cease to be such for our successors. From the age of the four so-called "classical elements" (none of them elements for us), down to comparatively recent times, the annals of science record many of these progressive simplifications; and we should be presumptuous to doubt the possibility of their recurrence hereafter.

LECTURE III.

Compound bodies—their volumetric constitution and condensation ratios as exemplified in hydrochloric acid—in water—in ammonia—illustrative experiments. Mechanical mixture and chemical combination—their distinguishing characteristics—experimental illustrations thereof—mixtures and combinations of the elements of hydrochloric acid—of water—constancy of chemical composition—marked changes of property attending chemical combination.

For the purpose of gaining additional points of view, we resume the study of the compounds, hydrochloric acid, water, and ammonia, which we have selected to inaugurate our studies; and we will proceed to examine, in the first place, the volumetric proportions in which the elements, hydrogen, chlorine, oxygen, and nitrogen combine to form these compounds.

In order to ascertain the volume ratio in which hydrogen and chlorine combine to form hydrochloric acid, we must repeat the decomposition of this compound, under conditions calculated to permit the measurement of the hydrogen evolved.

For this purpose we employ a U-shaped glass tube, about 50 centimetres long by 1·5 in diameter, having one sealed and one open limb; and we fix it upon a convenient stand. Just above the bend of the tube, its open limb has a small outlet tube (blown on at the lamp), and to this is affixed a piece of caoutchouc tubing, with an elastic wire nipper attached, by the action of which the caoutchouc tube is pinched close, but can be readily opened at pleasure. This arrangement, by-the-way, is often a convenient substitute for the ordinary stop-cock; and it is in such constant use in the laboratory that, to avoid periphrasis in referring to it, we will, if you please, agree to call it a *nipper-tap*. It is shown of full size in Fig. 43, side by side with another little arrangement of like kind, in which the com-

Fig. 43.

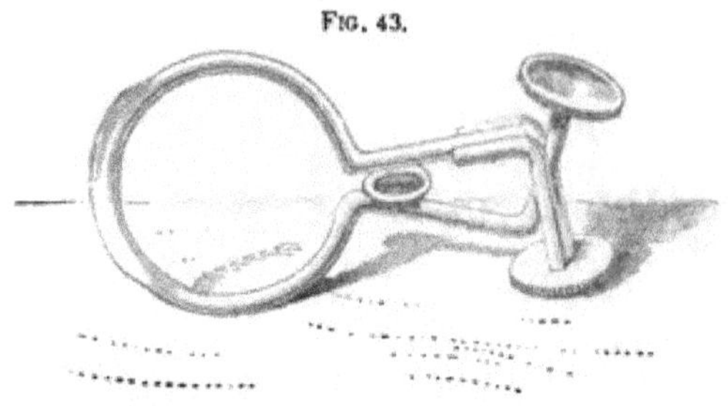

pression of the elastic tube is effected by means of a screw (Fig. 44). The use of the nipper-tap, or screw-tap, in the present case, is to facilitate the introduction of an appropriate volume of the gas to be examined into the apparatus. For this purpose the U-tube is first filled with mercury, and then the nipper-tap is set open, so as to afford a gradual exit to the metal in the open limb. The delivery-tube of an appropriate gas-generating apparatus is then passed down the open limb to the bend of the tube, in such manner that the gas bubbles up through the mercury into the sealed limb, from which, of course, the mercury escapes as the gas enters, volume for volume. An appropriate quantity of dry hydrochloric acid gas having been thus introduced, the nipper-tap is closed, and mercury is poured into the apparatus, until it stands at the same level in both limbs. The space occupied in the tube by the gas is then marked in any convenient way; preferably by a caoutchouc ring slipped over the tube (Fig. 45).

Fig. 44.

That portion of the open limb which is unoccupied by mercury is then filled with sodium-amalgam (Comp. p. 9), and the orifice of the tube is closed, either by the thumb or by a glass stopper. The gas may now, by inclining the tube adroitly, be easily transferred from the sealed to the stoppered limb; traversing of course, in its passage, the column of sodium amalgam, and being thereby decomposed. To insure complete decomposition, the apparatus should be once or twice shaken, so as to bring every portion of the gas into thorough contact with the amalgam; after which, by reversing the previous inclination of the tube, the gas may be re-transferred to the sealed limb of the apparatus. On removing the stopper or thumb from the mouth of the open limb, the mercury falls a little therein, and may be further lowered by opening the nipper-tap. As soon as the mercury stands at a uniform level in the two limbs, the gas is found reduced to exactly half its original volume (Fig. 46). The gas which remains, it need scarcely be remarked, is hydrogen, readily recognizable by its inflammability.

Fig. 45.

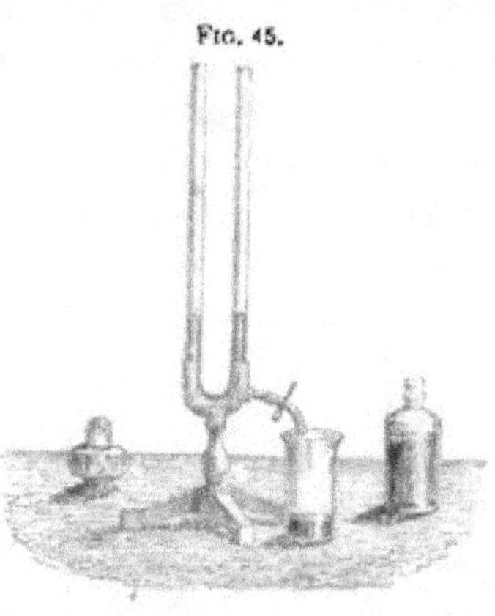

Fig. 46.

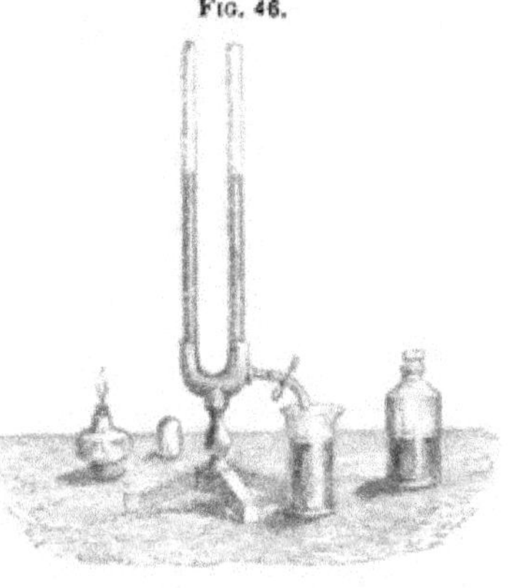

This experiment shows us that any given bulk of hydrochloric acid contains half that bulk of hydrogen. It only, therefore, remains to determine with what proportion, by volume, of chlorine this bulk of hydrogen is combined in hydrochloric acid.

This we learn from a second experiment. We again submit hydrochloric acid to electrolysis in the apparatus, and with the results already described (comp. p. 15). Hydrogen is copiously evolved at the negative pole, while chlorine is simultaneously liberated at the positive pole. At starting, however, the chlorine is almost entirely absorbed by solution in the surrounding liquid: nor is it till this is saturated that the chlorine begins to be manifested in a stream of bubbles, like those which, from the first, mark the escape of hydrogen at the opposite pole. At this stage of the process, the delivery-tube of the apparatus is attached, by means of a caoutchouc connector, to a glass tube, of about 40 or 50 centimetres long by 1·5 centimetre in diameter, drawn out, before the lamp, to a fine point at each end. This tube is then filled with the mixture of hydrogen and chlorine evolved by the electrolysis of hydrochloric acid. In order to expel every trace of air it is necessary that the mixed gases should be suffered to traverse the tube for a considerable time, and, as chlorine is a very noxious gas, it is necessary to take means for preventing its escape into the air. For this purpose, the free end of the tube is connected with the lower part of an upright cylinder containing coke, moistened with an alkaline liquid capable of absorbing the chlorine. Through this apparatus, which is termed a coke column, or coke tower, the gas streams up, parting with its chlorine as it ascends; so that only the hydrogen, which is innocuous, escapes. After the lapse of one or two hours the operation may be considered complete. The tube being now detached, its fine-drawn ends are immediately sealed.

This manipulation, by-the-way, requires considerable precaution, when performed upon explosive mixtures such as that with which we have now to deal. The object to be attained is the prevention of contact between the flame and the explosive gas during the sealing of the tube. Such contact would inevitably take place were the flame applied to seal the tube at its extreme tip or orifice. To avoid such contact—in other words, to keep a layer of glass interposed between the gas and the flame—this latter should be directed, not upon the orifice, but on the long drawn neck between the orifice and the body of the tube. This neck, being slender, soon softens; and its sides, running together, obliterate the channel, and so seal the tube. The excess of neck beyond the point thus sealed may then, while the glass is still soft, be detached, so as to leave a neatly-formed extremity. Thus conducted, on a neck sufficiently drawn out and attenuated, the sealing operation is tolerably safe; though, to guard against injury from accidental explosion, through the overheating or cracking the glass during the process, it is prudent to envelop the body of the tube in a cloth.

The tube having been sealed at each end, its gaseous contents have next to be examined. For this purpose it is requisite to bring the mixed gases into contact with a fluid capable of absorbing the chlorine but not the hydrogen. Water answers this purpose, and if a little soda be mixed with it the absorptive power is increased. Again, the addition of a vegetal colour—of an infusion of logwood, for instance—to tint the soda-solution employed, is useful as a means of evincing the presence of chlorine by exhibiting its bleaching action on the colour, so soon as it comes into contact therewith. By plunging the sealed finely-drawn extremity of the tube into a solution so prepared, and then breaking it off, the desired contact is effected, absorption begins, and the liquid is seen slowly rising into the tube to occupy the space vacated by the absorbed chlorine. This absorption goes on very slowly, however, because of the extreme minuteness of the surface of fluid exposed to the gas in the finely-drawn tube represented by the broken extremity. A great acceleration would evidently be obtained if the surface of contact could be extended; if, for example, we could wet the whole interior surface of the tube with the absorptive liquor.

Fig. 47.

We have here a contrivance for accomplishing this object (Fig. 47).

It consists of a caoutchouc connector tightly fitted to the end of the tube, so as to cover and enclose its fine-drawn sealed neck. This connector is provided with a small glass funnel, through which it can be filled with a tinted solution of soda, and has also a stop-cock, by which, when so filled, it can be closed. These arrangements being made, the fine-drawn neck becomes immersed in the solution; so that, on breaking it (Fig. 48), which the flexibility of the connector allows to be easily done, the solution finds its way through the orifice into the interior of the apparatus. By suitably inclining this,

Fig. 48.

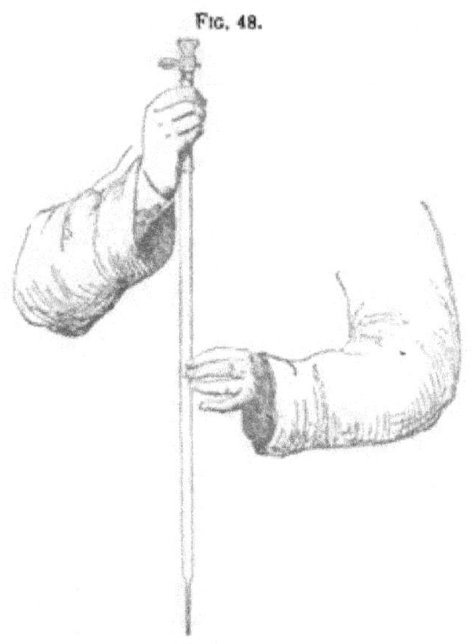

the solution may be caused to extend in a film over its interior, so as to expose a widely-spread surface to the gaseous mixture. The absorption of the chlorine is thus made to proceed with greatly increased rapidity, as is evinced by the speedy ascent of a small volume of the liquor into the wide part or body of the tube. This accomplished, that end of the tube which is armed with the funnel may be plunged under water, the flexible connector withdrawn, and the experiment continued in the ordinary way, by allowing the absorption to proceed, and the column of liquor to ascend in the tube till all the chlorine is dissolved. This is known to have taken place by the liquor's ceasing to rise in the tube.

The tube is now to be more deeply immersed in the liquor (the receptacle for which should be a tall glass cylinder, to facilitate this part of the manipulation), until the water-level within and without the tube is brought into coincidence. It is found that the tube is just half filled with liquor—in other words, that just half its gaseous contents have been absorbed. That the absorbed gas is chlorine is readily proved by the bleaching effect exerted by it on the logwood solution.

The nature of the residuary gas is as readily demonstrated by immersing the tube more deeply in the surrounding liquor, then breaking off its upper finely-drawn point and applying to the jet of gas, thus forced out by water pressure, a lighted taper; when it immediately takes fire, and burns with the characteristic pale-blue flame of hydrogen.

These phenomena furnish a simple and satisfactory reply to the question left unanswered by our previous experiment.

The action of sodium upon hydrochloric acid taught us that 2 volumes of hydrochloric acid contain 1 volume of hydrogen; the electrolysis of hydrochloric acid proves that, to form this compound, 1 volume of hydrogen combines with 1 volume of chlorine.

The two experiments, taken together, supply us with the exact points of information which our previous investigation of hydrochloric acid left deficient; so that, summing up our previous and present results, we now possess a complete and irrefragable demonstration, first, that hydrochloric acid is *composed* of hydrogen and chlorine; secondly, that these two elements are its *sole* constituents; thirdly, that they are united in *equal volumes* to form it; and lastly, that, in so uniting, they undergo *no condensation*, but produce a volume of compound gas, equal to the sum of the volumes of its elementary constituents.

This last-mentioned fact—the union of hydrogen and chlorine without contraction or expansion—may be illustrated by another, and an equally conclusive, experiment. While the electrolytic apparatus, used in the experiment just made, is still evolving hydrogen and chlorine in the proportion in which the two gases exist in hydrochloric acid, we may replace the wide glass tube, previously used, by another tube of equal length, but of stouter glass, and of smaller bore; half a centimetre being a convenient diameter. The two ends of this tube are, like those of the tube used in the previous experiment, drawn out into very fine necks. So soon as the tube is thoroughly purged of air, and exclusively filled with the gaseous constituents of hydrochloric acid, its fine necks are sealed by the blow-pipe jet, and its contents are exposed to the action of light for the purpose of inducing the combination of the mixed gases.

This curious effect may be obtained either by natural or artificial light. The direct rays of the sun produce instantaneous combination. But as such rays are not at our command in all seasons and at all places, as, for instance, during the earlier weeks of a London winter, it is convenient to know of an artificial light sufficiently intense to bring about the same effect. Such a light is that of the blue flame evolved by the combustion of two substances called nitric oxide and bisulphide of carbon; the properties of which will claim our attention hereafter. At present we need only note the simple manipulation needful for generating this light.

For this purpose some 8 or 10 cubic centimetres of the fluid called bisulphide of carbon are introduced into a tall glass cylinder filled with nitric oxide gas. This is most conveniently accomplished by means of a very thin bulb of glass, blown to the required size, filled with the bisulphide, and then sealed at the lamp. The glass cover of the vessel, already filled with nitric oxide gas, is drawn aside, the bulb dropped in, and the cover quickly replaced. Contact of atmospheric air is thus almost entirely obviated. The vessel is then jolted, so as to break the glass bulb, and the desired mixture of gas and vapour is at once obtained. A match is now applied to the opened mouth of the cylinder, when the mixture within takes fire, and burns with a brilliant, intensely blue flame, which descends into the vessel. The radiance of this light instantaneously induces the combination of hydrogen and chlorine; the effect being indicated by a flash of light, accompanied by a slight clicking sound, and followed immediately by the disappearance of the greenish colour of the mixture.

The figure (49) shows the disposition of the apparatus. To the left is the glass cylinder in which the light is generated: to the right are placed the mixed gases to be acted on (two tubes, instead of one being filled therewith, and employed in the experiment, to afford a double chance of success—this being an illustration which occasionally fails from causes not perfectly ascertained.

FIG. 49.

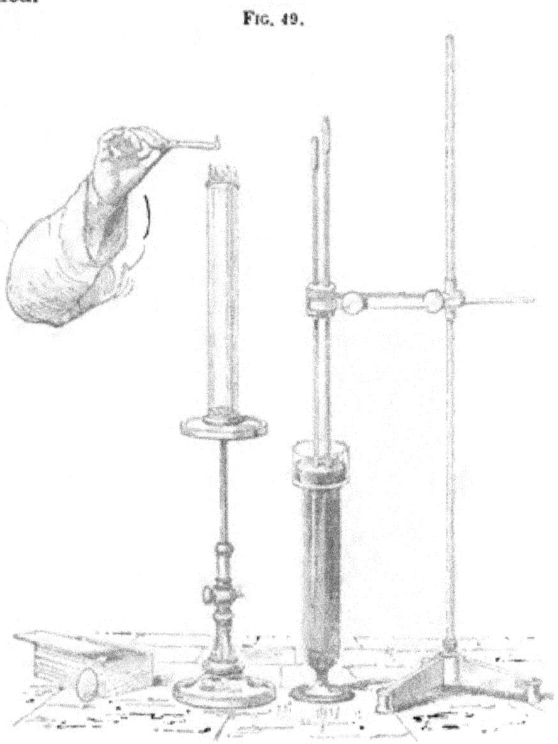

For the purpose of examining the product, one point of the tube is broken under mercury; when the first indication is immediately obtained. It is observed that neither does gas escape from, nor mercury penetrate into, the tube; from which it is clear that the combination of the gases has taken place without either contraction or expansion of their volume.

The next indication is obtained by pouring water on the mercury, and raising the tube so that its orifice, instead of plunging into mercury, may open into water. A striking effect is now produced. The state of equilibrium which obtained so long as the tube opened into mercury instantly ceases. The water no sooner comes into contact with the gas than this latter is dissolved; and so rapid is the absorption of the gas by the water that this rises in the tube, filling it almost instantaneously. The solution thus obtained is readily shown, by the application of appropriate tests as before, to be dilute hydrochloric acid; and these results confirm our previous experimental proofs that hydrochloric acid is formed by the union of hydrogen and chlorine gases, in equal volumes, without condensation.

In performing this experiment, whether sunlight or the light of bisulphide of carbon be employed, some manipulatory precautions are necessary to protect the operator from possible injury. The mixed gases in the act of combination evolve so much heat, and are so powerfully dilated thereby, as sometimes to burst the tube containing them. The experimentalist should, therefore, not omit to protect himself by a screen, for which purpose a sheet of stout plate-glass may be conveniently employed. Thus, even should the tube explode, the dangerous scattering of its fragments is prevented. It is, however, only rarely that the body of the tube is shattered: in most cases the fracture is confined to one or other of the sealed points. To avoid the loss of the experiment by an accident of this kind, the upper point of the tube may be strengthened by imbedding it in sealing-wax, which may be most conveniently applied by fusing a little in a small piece of glass tube (see Fig. 49), sealed at one end, and plunging the point to be protected into the fused mass, which is then allowed to cool and harden. As for the lower point, escape of gas from this, in case of rupture, is readily obviated by keeping it plunged into a trough-cylinder filled with mercury.

The ratio in which hydrogen and oxygen are associated in water is most conveniently established by the electrolysis of this compound.

We have already availed ourselves of the action of the electric current for the purpose of ascertaining the nature of the constituents of water; and we learned incidentally that the hydrogen was evolved in larger measure than the oxygen. An appropriate modification of the apparatus then used enables us to determine with precision the ratio in which they are evolved.

For this purpose, two glass tubes of equal diameter, each sealed at one end and open at the other, are filled with acidulated water, and suspended, mouth downward, over a basin also filled with acidulated water, below the level of which the mouths of the two tubes dip (Fig. 50). The conducting-wires, or poles of

FIG. 50.

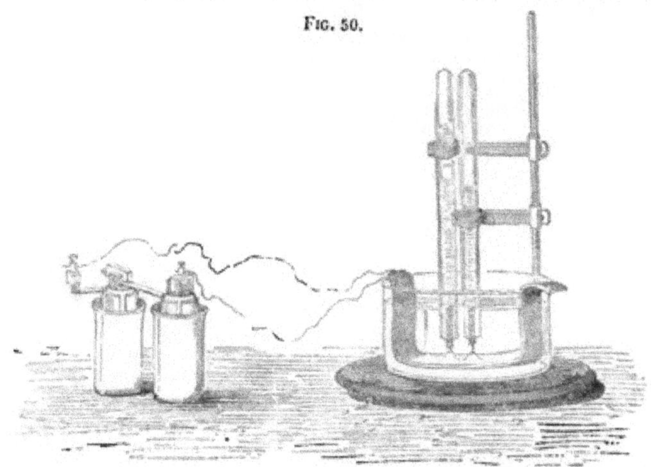

the battery, duly insulated by a gutta-percha covering (except at their points), are immersed in the fluid of the basin, and their points are disposed one below each tube. Each point is armed with a small strip of platinum-foil.

The battery is now put into action, and its current, transmitted by the platinum poles through the water which lies between them, decomposes this into its two constituent gases, which appear, as before, ascending in bubble-streams, one from the positive, the other from the negative pole: this latter stream being (also as before) manifestly the more abundant of the two.

As the two gases collect in the respective tubes, it becomes obvious that exactly two volumes of the more abundant are evolved for every one of the less abundant constituent. Now, we know from former experiments that the more abundant gas (that disengaged at the negative pole) is hydrogen, whilst the less abundant product (that liberated at the positive pole) is oxygen. This we can, of course, again verify as before, by removing the tubes, and applying to their contents the tests with which we are already acquainted. It is thus proved that in water two volumes of hydrogen are united with one volume of oxygen.

Fig. 51.

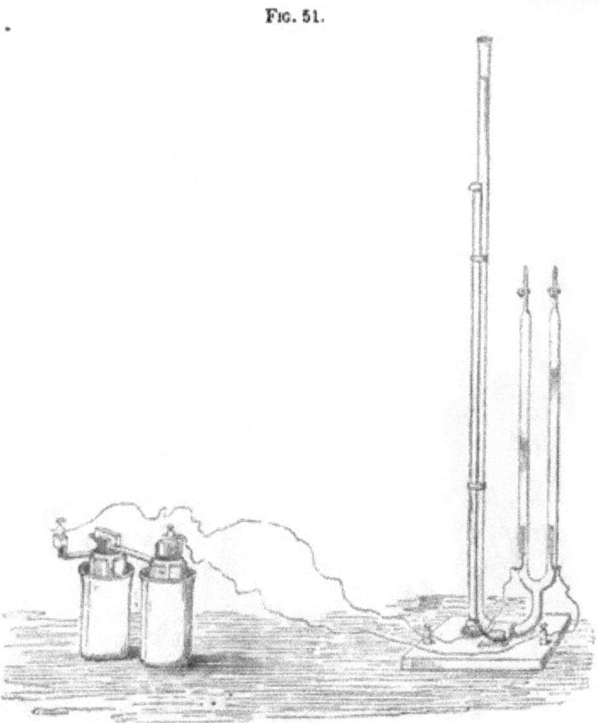

Fig. 51 shows an improved form of apparatus for demonstrating this important fact. Instead of the two sealed tubes separately inverted in a basin, as just described, we have here a three-branched tube, with one long and two short limbs. The long limb acts as a water-reservoir, instead of the basin; the two short limbs, which are fitted with stop-cocks, or nipper-taps, above, and which communicate freely with the long limb below, contain, intermediately, each a platinum electrode. When this apparatus is charged with water, and the electric current is transmitted through the electrodes, the two gaseous products are evolved, each at its appropriate pole, so as to accumulate separately, each in the corresponding limb. As this accumulation proceeds, the water in the short limbs is forced downward out of these, so as to rise in the long limb, forming a column, the weight of which serves, in the sequel, to expel the gas from each short limb through the corresponding stop-cock, when this is opened for the purpose of testing the nature of the gas obtained.

It remains to be investigated whether hydrogen and oxygen, during their conversion into water, undergo any change of volume; or, like hydrogen and chlorine, combine to produce a measure of compound gas, exactly equal to the joint measure of the constituents.

To determine this point it is necessary to compare the volume of the elementary water-constituents with that of the water formed, at a temperature high enough to maintain the latter in the purely gaseous condition in which engineers term it *dry steam.*

Fig. 52.

The experiment is made in a U tube similar to that used in analyzing hydrochloric acid (p. 41). The closed limb of the tube is fitted, at a point near its sealed extremity, with two platinum wires, welded to the glass, which they traverse, and separated at their inner ends by a distance of only two millimetres; while their outer ends are formed into loops for the attachment of appropriate battery-conducting wires.

This arrangement is shown, upon an enlarged scale, in the figure (52), in which two slightly-modified forms of it are placed side by side; the only difference being that in one the platinum points are bent slightly up within the tube, to facilitate the cleansing of its interior.

The general disposition of the apparatus is depicted in Fig. 53, which shows the electric arrangements to the left; and the U tube, with its adjuncts, to the right. The so-called "induction coil," interposed in the electric circuit causes the current to pass in the form of a stream of sparks between the wires within the U tube. Into the sealed limb of this tube, which is filled with mercury, we admit a column, about 25 or 30 centimetres high, of a mixture of hydrogen and oxygen in the proportions in which they form water. This mixture may of course be obtained by adding two volumes of hydrogen to one of oxygen; it is, however, prepared much more readily in due volumetric proportion, and in a state of perfect purity, by the electrolysis of water, as already described. The gas-filled limb of the U tube is surrounded by a high glass cylinder, the lower mouth of which is fastened around it

Fig. 53.

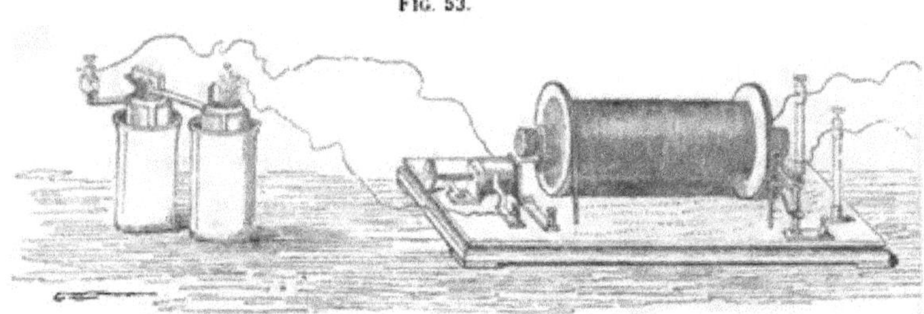

by means of a perforated cork, whilst its upper mouth (likewise closed by a cork) rises about five centimetres above the sealed extremity. The annular space thus formed communicates, at its upper end, by means of a bent glass tube and a perforated cork, with a flask which contains a liquid having a boiling point considerably above that of water: amylic alcohol, which boils at 132° C., is well adapted for the experiment. On protracted ebullition, the vapour descends from the flask into the annular space, which rapidly acquires a uniform temperature of 132°. To prevent the powerfully-odoriferous vapours from escaping into the atmosphere, the lower extremity of the glass cylinder is connected with an appropriate vapour-condenser, such as a glass tube kept cool by a current of water. Under the influence of heat, the column of mixed hydrogen and oxygen in the tube expands, and its height is marked by any suitable means; preferably by slipping a caoutchouc ring over the outer glass cylinder. Care must of course be taken, before doing this, to bring the mercury to a uniform level in both limbs of the U tube, either by adding or withdrawing a proportion of the metal as required. A little more mercury is then poured into the open limb, which is, lastly, closed by a well-fitting cork. Between this cork and the

Fig. 53.

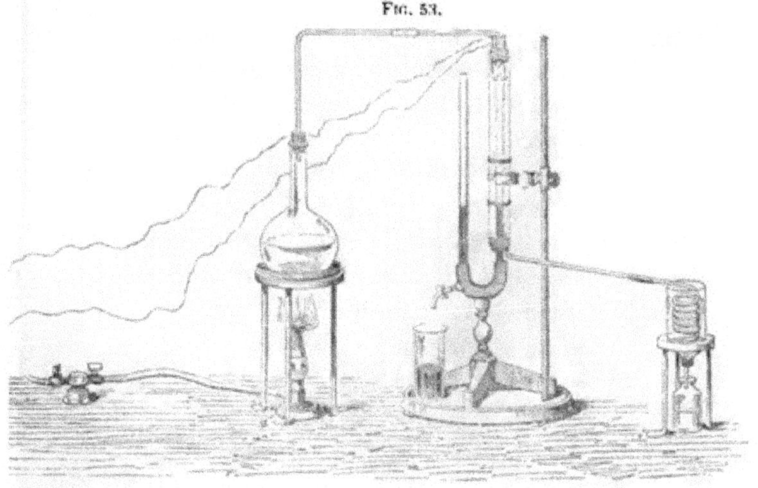

mercury intervenes a column of air, some eight or ten centimetres in length, and capable of yielding to pressure, like a spring. It now only remains to inflame the gaseous mixture by causing the current of the induction-coil to leap, in the form of a spark, between the platinum points. The gases combine with an explosion; which is, however, much mitigated in violence by the elastic action of the above-mentioned air column. At the high temperature employed (132°), the water formed retains the gaseous condition. On removing the cork, and allowing the mercury to flow through the nipper-tap, until it is level in both limbs of the U tube, it becomes obvious that the original measure of mixed gases is diminished by one-third; the residuary two-thirds are water-gas, or dry steam, which condenses into liquid water so soon as the tube is allowed to cool.

Thus, therefore, it stands experimentally demonstrated, first, that hydrogen and oxygen undergo condensation in combining to form water; and, secondly, that the volume of the water-gas produced holds an extremely simple ratio to the volume of its constituent gases; two volumes of hydrogen and one volume of oxygen condensing, by their union, into two volumes of water-gas.

The method of ascertaining the volume-ratio in which hydrogen and nitrogen combine to form ammonia is less simple than that which suffices for the corresponding study of hydrochloric acid and water.

For this purpose chlorine is employed, as before, to withdraw hydrogen from ammonia, and set free nitrogen; and besides this, means are adopted to determine with accuracy the volume of nitrogen thus separated from a known quantity of ammonia.

A glass tube for holding chlorine, and a globe for receiving solution of ammonia, and admitting it, drop by drop, to the chlorine, constitute the requisite apparatus. The glass tube is from 1 to 1·5 metre long, sealed at one end, open at the other, and marked off, by elastic caoutchouc rings slipped over it and clipping it firmly, into three equal portions. The globe has a stoppered aperture above, and a dropping tube (Fig. 54) drawn out to a narrow orifice below. This tube is fitted with a stop-cock, and passes through a perforated cork, by means of which it can be tightly fixed into the open mouth of the chlorine-tube.

Fig. 54.

The apparatus is thus employed. The long chlorine-tube having been filled with cold water and inverted over a pneumatic trough, with its mouth immersed below the water-level, is filled with chlorine gas in the usual way. When full, it is still allowed to stand for about fifteen minutes over the chlorine delivery-tube, that its interior surface may be quite freed from the chlorine-saturated water that else would remain adherent thereto. The globe, meanwhile, is filled with a strong solution of ammonia, and its stop-cock is turned so that its dropping-tube also may be filled to its very tip with this solution. The globe is then stoppered, and the cock closed again; after which it is ready for connection with the chlorine-tube. To effect this connection without admission of air into the chlorine-tube requires some little care and dexterity. The globe has to be immersed in the pneumatic trough, with its dropping-tube upward, and in this position to be brought beneath the mouth of the chlorine-tube, into which the globe-tube is inserted, and fixed firmly by means of the cork which it carries. In effecting this junction, great care must be taken not to introduce any water from the trough into the chlorine-tube. This tube, with its ammonia-globe joined to it, may now be removed from the trough, and supported in a vertical position, with the globe surmounting it. A single drop of the ammonia-solution is now suffered to fall from the globe into the chlorine-tube, the stop-cock being opened for a moment for this purpose (Fig. 55). The entrance of this drop into the atmosphere of chlorine is marked by a small, lambent, yellowish-green flame at the drawn-out point of the dropping-tube. Drop by drop, at intervals of a few seconds, the ammonia-solution is allowed to fall into the chlorine-tube, the ammonia of each drop being at

FIG. 55.

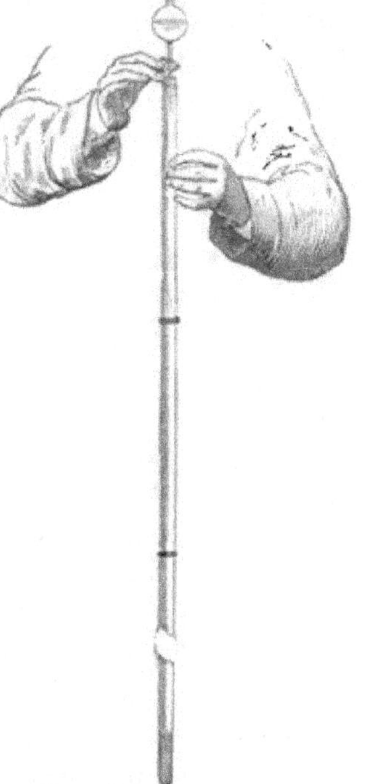

the instant of its contact with the chlorine converted, with a flash of light and the formation of a dense white cloud, into hydrochloric acid and nitrogen. The addition of ammonia must be continued till the whole of the chlorine present is supplied with hydrogen at the expense of ammonia. To insure this the ammoniacal solution is added in excess, a column of three or four centimetres being abundantly sufficient. The result is that the hydrochloric acid formed combines with the excess of ammonia to form a compound, of which we shall have to speak hereafter; but of which it is enough to say here that it makes its appearance as a white deposit, lining the interior of the chlorine-tube. This deposit, being soluble, is readily washed down and dissolved by agitating the liquor in the tube, which now contains the whole of the nitrogen separated, except a little which remains dissolved in the liquor. This small quantity of dissolved nitrogen is easily expelled from the liquor by heat.

We are now sure of two points; viz., that the whole of the chlorine has been converted into hydrochloric acid at the expense of the ammonia; and, secondly, that we possess within our tube the whole of the nitrogen thus set free.

It becomes our next object to withdraw the excess of ammonia. For this purpose dilute sulphuric acid, which fixes ammonia, is introduced by means of the globe previously employed to admit ammonia.

The nitrogen, being thus freed from all intermixed gaseous bodies, has only now to be brought to mean atmospheric temperature and pressure in order that it may be ready for measurement.

The temperature, which had been raised by the application of heat to the liquor to expel the dissolved nitrogen therefrom, is readily brought back to the mean by plunging the tube into cold water. To equalize the pressure within and without the tube, the bent syphon tube (Fig. 56) is employed. One end of this communicates with the interior of the tube, while the other plunges beneath the surface of water subject to atmospheric pressure. That this pressure exceeds that of the gas in the tube is at once seen by the flow of the water through the syphon into the tube. As the water-level in the tube rises, the nitrogen, previously expanded, gradually approaches its normal volume, which it exactly attains when the flow ceases, showing the pressure within and without to be in equilibrium. Both temperature and pressure being now at the mean, all the requisite conditions are fulfilled for obtaining an exact knowledge of the true volume of nitrogen; and this, on inspection, is found exactly to fill one of the three divisions marked off at the outset on our tube.

Fig. 58.

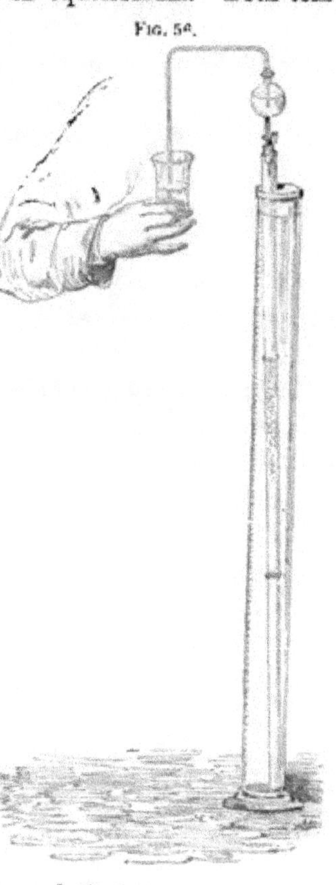

Now, bearing in mind that we started with the three divisions full of chlorine, and that we have saturated this chlorine with hydrogen supplied by the ammonia; bearing in mind, moreover, that hydrogen combines with chlorine, bulk for bulk; it is evident that the one measure of nitrogen which remains in the tube has resulted from the decomposition of a quantity of ammonia containing three measures of hydrogen.

It is, therefore, clearly proved by this experiment that ammonia is formed by the union of three volumes of hydrogen with one volume of nitrogen.

Thus much determined, it remains to ascertain the condensation undergone by these elements in combining to form ammonia.

Of this we may obtain ocular demonstration by a very simple experiment. We cannot, indeed, employ in the case of ammonia the method adopted to determine this ratio in the cases of hydrochloric acid and water; we cannot, that is to say, mix hydrogen and nitrogen in due volumetric proportion to form ammonia, and then cause them to unite, and measure the space occupied by the product. The direct synthesis of ammonia has never yet been accomplished; so that we must fall back upon analysis to furnish us with this demonstration. We must split up a measured quantity of ammonia into its constituents, and compare the space occupied by the ammonia before treatment with the space filled by its separated constituents. This we are enabled to do very easily, by availing ourselves of the tendency of ammonia to break up into its elements under the influence of a moderate heat.

For this purpose we need our often-employed U tube as a receptacle, and the spark-stream supplied by the electric current in traversing our induction-coil as a source of heat (Fig. 57).

The sealed limb of the U tube is filled to about one-third of its height with dry ammonia, over mercury, and the height of the column of gas is accurately measured; care having been taken, as usual, to bring the mercury in each limb of the tube to a uniform level. The spark-stream is now set flowing between the platinum points, and the volume of the gas on which it acts is immediately observed to increase. This dilatation continues for some five or ten minutes (according to the quantity of ammonia under treatment); and, when it ceases, the level of the mercury (disturbed, of course, by the expansion of

FIG. 57.

the gas) is readjusted to perfect uniformity in both limbs of the tube; when it is immediately perceived that the original volume of gas has become doubled. If a little of the gas (previously so pungent) be allowed to escape from the tube, by a stop-cock provided for that purpose, it is found to have become inodorous; while the presence of hydrogen is indicated by its inflammation on the approach of a light.

This experiment proves that hydrogen and nitrogen, as combined in ammonia, occupy only half the space they fill in their free state; or, in other words, that 4 volumes of the mixed gaseous constituents of ammonia, composed, as we have already ascertained, of 3 volumes of hydrogen and 1 volume of nitrogen, condense, during their combination, to form 2 volumes of ammonia.

The result of these inquiries into the composition of hydrochloric acid, water, and ammonia may be summed up as follows:—

1 vol. of hydrogen + 1 vol. of chlorine = 2 vols. of hydrochloric acid.
2 vols. of „ + 1 vol. of oxygen = 2 vols. of water gas.
3 vols. of „ + 1 vol. of nitrogen = 2 vols. of ammonia.

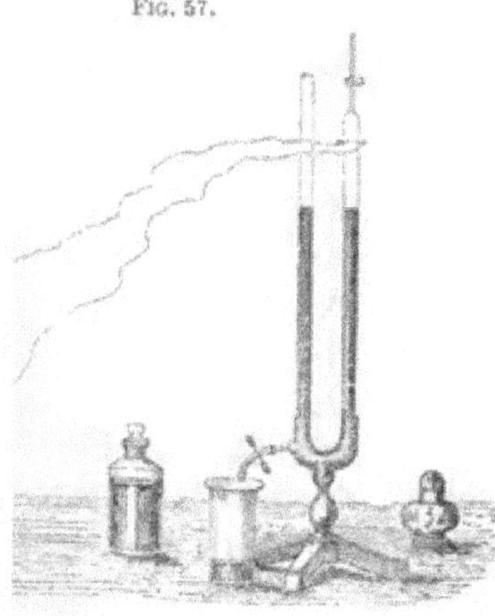
FIG. 57.

Thus it is evident that hydrochloric acid, water, and ammonia, not only differ as to the volume-ratio of their elementary constituents, but also as to the ratios of the spaces occupied by these *before* and *after* their combination to form chemical compounds. These ratios rise from unity, or $\frac{1}{1}$, in the case of hydrochloric acid, through $\frac{2}{3}$ in the case of water-gas, to $\frac{1}{2}$ in the case of ammonia; the condensation increasing, in these cases, *pari passu* with the complexity of the chemical compound.

Notwithstanding these remarkable differences in their chemical construction, the three compounds, hydrochloric acid, water, and ammonia, illustrate several important general laws, which deserve our particular attention. Among the most striking of these are:—1, the immutability of the proportions in which the elements are in each case associated; and, 2, the strongly-marked changes of character and properties induced by the union of the elements to form these compounds respectively. Hydrogen and chlorine, hydrogen and oxygen, hydrogen and nitrogen, may be *mixed* in any proportion whatever, and their respective properties still remain traceable in the mixtures; but in the *compounds* of hydrogen with chlorine, of hydrogen with oxygen, of hydrogen with nitrogen, designated respectively hydrochloric acid, water, and ammonia, the elements are *combined* only in the determinate proportions, and undergo only the determinate changes of property, which we have experimentally verified.

This constancy of composition and character in chemical compounds is a law to which there is no exception.

The immutability of the ratios in which the elements unite to produce compound bodies is strikingly illustrated by the formation of hydrochloric acid and water.

To demonstrate this fact, we will employ a glass tube, divided by a glass stop-cock into two compartments of unequal capacity, and having its open ends provided with glass stoppers (Fig. 58). With this apparatus we will make two experiments.

In the first experiment we will fill the smaller compartment, which is only about half the size of the larger, with dry hydrogen, the larger, with dry chlorine; and after bringing the two into communication by turning the stop-cock, we will expose the apparatus to diffuse daylight. The two gases gradually commingle, and soon begin to combine. The reaction is completed by exposing the tube for a short time to the influence of the *direct* solar beams. On opening the tube under water, the liquid is found to rise and fill a space double that of the smaller, or hydrogen compartment. The residuary gas is chlorine; the volume of which, owing to its solubility in water, rapidly diminishes.

In the second experiment we will fill the smaller compartment with chlorine, the larger one with hydrogen, and allow the two gases to act upon one another under the influence of light, as before. The volume of water which now rises in the apparatus is equal to that previously observed; but the residuary gas is now hydrogen.

In these contrasted experiments, hydrogen and chlorine, though mixed in reversed proportions, have nevertheless combined in volumes absolutely unaltered; the excess, whether of the one or the other gas, taking no part in the reaction.

Fig. 58.

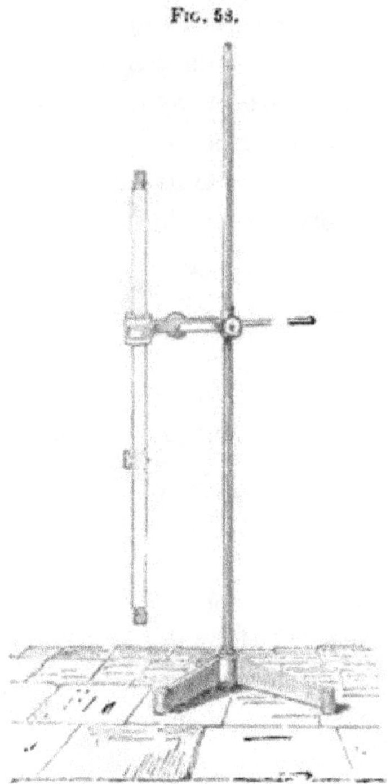

Analogous phenomena are witnessed during the formation of water by the combination of hydrogen with oxygen.

To prove this experimentally, three U tubes, provided with spark-wires and outlet-pipes, similar to those previously employed (comp. p. 51), are filled with mercury, and fixed in an appropriate stand (Fig. 59). By the electrolysis of water in the apparatus already described (comp. p. 24), we obtain a mixture of hydrogen and oxygen gases in the ratio in which they exist in water. A column of this gaseous mixture, eight or nine centimetres high, is allowed to rise in each of the three tubes; the height being marked by caoutchouc clipping-rings, after the mercury has been rendered level in the two limbs of each apparatus by the usual means. Into one of the tubes we pass, moreover, an additional volume of hydrogen, into the other an additional volume of oxygen; these additional volumes being likewise marked by means of caoutchouc rings.

Of the three U tubes, one—in our experiment the middle one—now contains a mixture of hydrogen and oxygen in the proportion in which the two gases are separated in the electrolysis of water; the other contains the same mixture + an additional volume of hydrogen; the third contains the same

Fig. 59.

mixture + an additional volume of oxygen. The spark of the induction-coil is now passed through the three mixtures, giving rise in each of them to the formation of water. In the middle tube every trace of gas is now found to have disappeared, the mercury filling the tube to the very top; in the second and third tubes, volumes of gases remain, which, after appropriate adjustment of the mercury in the two limbs, are found to be equal respectively to the volume added in each case; and these gases, when liberated through stop-cocks adapted to the tubes for that purpose, are respectively recognized, by means of the usual tests, as hydrogen and oxygen.

Hence it appears that, in these three experiments, hydrogen and oxygen did exclusively combine in the proportion in which the two gases had been originally evolved from water. The hydrogen and oxygen afterwards respectively admitted in excess took no part in the formation of water.

The essential changes which the properties of the elements undergo in associating to form chemical compounds are sufficiently obvious from experiments with which we are already familiar.

It would be impossible to quote a more instructive example

FIG. 59.

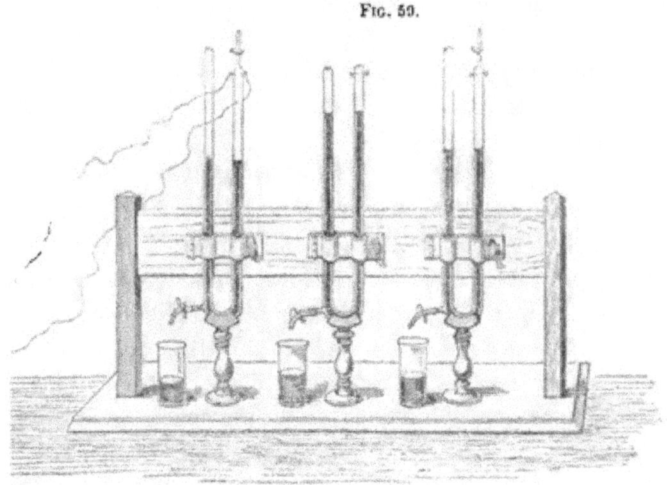

of such changes than that furnished by the transformation of a mixture of hydrogen and chlorine into hydrochloric acid. In the mixture of the two gases obtained by the electrolysis of hydrochloric acid, the properties of each gas are easily recognized. The mixture retains the fundamental property of hydrogen; it is still inflammable. The mixture also possesses the yellowish-green colour of chlorine, its odour, and its bleaching powers. On treating the mixture with water, its soluble constituent is absorbed, its colour, odour, and bleaching powers become weaker and weaker, until at last there only remains the colourless, inodorous, tasteless hydrogen. Now let the mixture, by one of the processes with which we have become acquainted, be converted into a chemical compound, and instead of the yellowish-green, colour-bleaching, suffocating chlorine, sparingly soluble in water—instead of the inodorous, tasteless, inflammable hydrogen—we have a colourless gas, possessing no longer the slightest bleaching power, absorbed by water with avidity, of pungent odour and taste, and utterly incapable of combustion.

In like manner, even a cursory comparison of the properties of hydrogen and oxygen gases with those of the liquid water which they produce by their union exhibits differences as striking as can well be conceived; but even in water-gas (dry steam), the fundamental property of hydrogen (inflammability), and that of oxygen (power of supporting combustion), are found to be entirely extinct.

Not less striking is the change in the properties of hydrogen and nitrogen, when associated to form ammonia. Two elementary gases, inodorous, insoluble in water, without action on vegetal colours, are converted by chemical combination into a gaseous compound, possessing a most penetrating odour, capable of restoring acid-reddened litmus-paper to its pristine blue colour, and so intensely and rapidly soluble in water, that the fluid rushes into a tube filled with ammonia, as into a vacuum.

The preceding remarks sufficiently mark the difference between *mechanical mixture* and *chemical combination.* In mechanical mixtures the elements are capable of interfusion in any proportion whatever; in chemical compounds they unite in definite immutable proportions, volumetric and ponderal. The mechanical mixture exhibits properties intermediate between those of its ingredients; in the chemical compound the properties of its constituents become extinct, their individuality being, so to speak, merged in the formation of a new body, with new properties.

The recognition of these marked differences between mere mechanical mixtures and definite chemical compounds very naturally leads us to examine the conditions under which the former become converted into the latter. In this case, too, the formation of hydrochloric acid and of water affords us welcome elucidations.

A mechanical mixture of hydrogen and chlorine, as furnished by the electrolysis of hydrochloric acid, may, if due care be taken to protect it from the action of light, be preserved for an indefinite period without undergoing the slightest change. Under the influence of ordinary diffuse daylight, the transformation of the mixture into a compound is accomplished, as we have seen, in the course of a few hours. Direct sunlight, or certain artificial lights of great intensity, effect the transition instantaneously (comp. p. 46). Combination then takes place with violent explosion, sometimes shattering the vessel which contained the mixture. Transformation of the mixture into a compound may in this case also be accomplished by contact with a burning body, or by the passage of an electric spark; both which processes are attended by explosive violence.

The transformation of a mechanical mixture of hydrogen and oxygen into the chemical compound, water, is less easily accomplished. It appears from recent experiments, that they also may be caused to combine by sunlight though only on very protracted exposure to its influence. On the approach of a flame, however, or the passage of the electric spark, the mixture instantaneously explodes, and is, at the same moment, converted into water.

Hence it appears that mechanical mixtures are often transformed into chemical compounds by the action of light, and, more frequently still, by the powerful influence of heat. The experience acquired in the study of hydrochloric acid and water cannot, unfortunately, be extended by the similar examination of ammonia, seeing that no process is known for directly effecting the transformation of a *mixture* of hydrogen and nitrogen into a *compound* thereof. But this exception renders it the more necessary to mention that the conditions which determine the formation of hydrochloric acid and water from mixtures of their elements suffice for the accomplishment of like results in an endless variety of cases; heat being especially distinguished as the most general promoter of chemical transformation.

When proceeding hereafter with the special study of the individual elements, we shall have frequently to resume the consideration of these conditions of chemical change. Opportunities will then also occur for examining, minutely, certain interesting concomitants of chemical activity; such as, for instance, the remarkable development of light and heat which we have already seen attending the reactions in several of our illustrative experiments.

LECTURE IV.

Chemical symbols — their nature and value — diagrammatic symbols — initialed—figured—symbolic equations constructed therewith—information thereby conveyed—formulæ thence derived—summary of information condensed in chemical formulæ—hydrochloric acid, water, and ammonia, considered as types of chemical combination—hydrobromic and hydriodic acids—their construction upon the type of hydrochloric acid—ponderal analysis and volume-weights of these compounds—volume-weights of bromine and iodine gases—sulphuretted and selenetted hydrogen—their construction upon the type of water—ponderal analysis and volume-weights of these compounds—volume-weights of sulphur and selenium gases.

The principal facts determined by experiment, as to the composition of hydrochloric acid, water, and ammonia, are susceptible of clear and concise expression in a few happily-chosen symbols, which experience has shown to be so powerful as instruments of chemical research, and so invaluable as adjuncts of chemical nomenclature, that they may justly claim precedence over all the subjects pressing for attention at this early stage of our inquiry.

If equal volumes of hydrogen, chlorine, oxygen and nitrogen (taken, of course, at like temperature and pressure) be represented by equal squares, having the initials of these elements inscribed therein, the composition, by volume, of hydrochloric acid, water, and ammonia, may be thus expressed :—

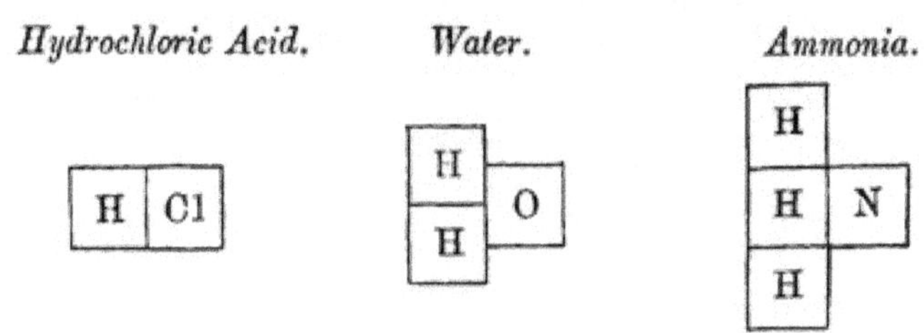

By inscribing within these symbolic squares the *volume-weights*

of the elements instead of the initial letters of their names, we obtain a series of *figured* symbols, which, when placed side by side with the *lettered* ones, furnish the following instructive equations, exhibiting on one hand the *volumes*, and on the other hand the *weights*, of the elements entering into combination :—

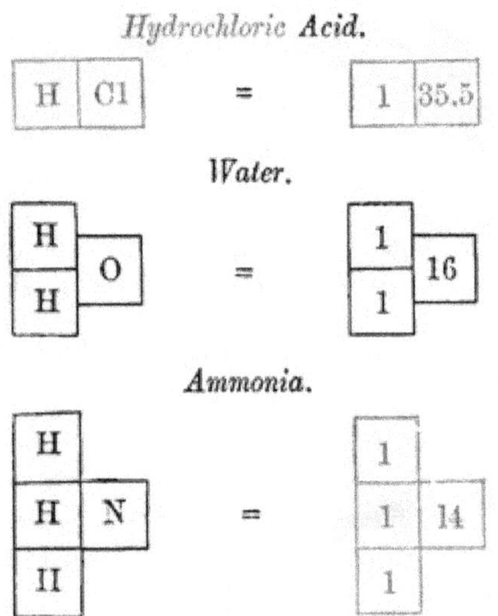

Again, if we add together the weights in the figured symbols respectively, thus :—

$$1 + 35{\cdot}5 = 36{\cdot}5$$
$$1 + 1 + 16 = 18$$
$$1 + 1 + 1 + 14 = 17$$

we learn the composition, by weight, of 36·5 parts of hydrochloric acid, of 18 parts of water, and of 17 parts of ammonia respectively. From these ratios, the centesimal composition of each compound is, of course, deducible by the well-known rule of proportion. Thus with respect to hydrochloric acid :—

$$36{\cdot}5 : 1 = 100 : x$$
$$x = \frac{100}{36{\cdot}5} = 2{\cdot}74 \text{ per cent. of hydrogen,}$$

and $100 - 2{\cdot}74 =$ $\quad 97{\cdot}26$ per cent. of chlorine,
forming $\quad \overline{100{\cdot}00}$ parts of hydrochloric acid.

So, again, with respect to water :—

$$18 : 2 = 100 : y$$

$$y = \frac{200}{18} = 11{\cdot}11 \text{ per cent. of hydrogen,}$$

and $100 - 11{\cdot}11 =$ $\quad 88{\cdot}89$ per cent. of oxygen,
forming $\quad \overline{100{\cdot}00}$ parts of water.

And so, lastly, with respect to ammonia :—

$$17 : 3 = 100 : z$$

$$z = \frac{300}{17} = 17{\cdot}64 \text{ per cent. of hydrogen,}$$

and $100 - 17{\cdot}64 =$ $\quad 82{\cdot}36$ per cent. of nitrogen,
forming $\quad \overline{100{\cdot}00}$ parts of ammonia.

By connecting, in our memory, the *volume-weights* of the elements with their respective *initials*, we have all the preceding information brought before us in the three simple symbolic formulæ given at the outset of this Lecture.

We must, however, proceed a step further, in order to obtain an equally clear and concise symbolic representation of the relative volume-weights (or specific gravities) of the three compounds under review—hydrogen being taken as unity.

For this purpose, we must bear in mind the experiments by which we established, firstly, that hydrogen and chlorine combine to form hydrochloric acid without condensation; secondly, that, during the union of hydrogen and oxygen, the volume of the water-gas formed is only two-thirds of the volume of the mixed gases; and thirdly, that the formation of ammonia involves the condensation of the constituents to one-half of their volume before combination. These facts may be graphically symbolized as follows :

Hydrochloric Acid.

H 1	+	Cl 35.5	=	HCl 36·5
1 vol.	+	1 vol.	=	2 vols.

Water.

H 1, H 1	+	O 16	=	H_2O 16
2 vols.	+	1 vol.	=	2 vols.

Ammonia.

H 1, H 1, H 1	+	N 14	=	H_3N 17
3 vols.	+	1 vol.	=	2 vols.

These symbolic equations show, at a glance, the fact (of which we have already had experimental proof) that very unequal weights of these compounds occupy, under the same conditions of temperature and pressure, exactly equal spaces; the space occupied, in each case, being exactly double the space filled by the unit-volume of hydrogen. In other words, the volume-weight of hydrogen being 1—

The double volume-weight of hydrochloric acid gas is			36·5
"	"	of water gas, or dry steam	18
"	"	of ammonia gas . .	17

Evidently, therefore, we have only to halve these double volume-weights of hydrochloric acid, water-gas, and ammonia, in order to arrive at their respective single volume-weights, relatively to hydrogen taken as unity; in other words, at their respective specific gravities: thus, 1 being the volume-weight of hydrogen, we have:—

$\frac{36 \cdot 5}{2} = 18 \cdot 25$ = the relative vol.-weight or sp. gr. of hydrochloric acid.

$\frac{18}{2} = 9$ = ,, ,, of water-gas.

$\frac{17}{2} = 8 \cdot 5$ = ,, ,, of ammonia.

When these ratios of volume, weight, and condensation, have been once firmly fixed in the memory by aid of the figured squares, this graphic or diagrammatic form of symbolism may be dispensed with, and the whole of our experimentally acquired knowledge of hydrochloric acid, water, and ammonia, may be embodied in the following succinct expressions:—

$$HCl = 2 \text{ vols.}$$

$$\left.\begin{matrix}H\\H\end{matrix}\right\}O \quad \text{or} \quad H_2O = 2 \text{ vols.}$$

$$\left.\begin{matrix}H\\H\\H\end{matrix}\right\}N \quad \text{or} \quad H_3N = 2 \text{ vols.}$$

The importance of these concise formulæ is best evinced by a simple recapitulation of the information they afford at a glance. We learn from them:—

1. The *number* and *names* of the elements entering into the composition of hydrochloric acid, water, and ammonia respectively.

2. The *ratios* in which the elements are united in these several compounds by *volume.*

3. The *ratios* in which the elements are united therein by *weight.*

4. The *ratios* in which the volumes of the several compounds, *when formed*, stand to the volumes of their respective constituents *before combination.*

5. The relative *volume-weights* or *specific gravities* of these several compounds in the state of gas or vapour, hydrogen being taken as unity.

The weight and volume ratios represented by the formulæ of hydrochloric acid, water, and ammonia,

$$HCl, \quad H_2O, \quad \text{and } H_3N,$$

are not merely interesting in themselves, but acquire a still deeper significance when it is found that they are types or models, representing as many groups of compounds, each cast, so to speak, in the same mould, and governed by the same structural law, as its generic prototype. These analogous bodies come next in order for consideration.

In our previous studies, devoted to the prototypes, we have adduced experimental proof of each one of our propositions. Similar evidence exists as to the structure of the analogues, which we have now to review. It would, however, encumber our present demonstration, and lead us too far from our immediate purpose, were we to attempt, at this stage of our progress, the experimental verification of every fact adduced in support or extension of our arguments. In the pursuit of scientific knowledge, much must at all times be admitted on adequate testimony; and we, like other students of nature, must refer for many of our facts to the recognized stores of science, reserving for a future period the experimental demonstration of some of the results in the meanwhile taken on trust.

Among the elementary bodies, which we shall hereafter have to investigate, there are two, named respectively Bromine and Iodine, whose properties, as we shall find, ally them closely with chlorine. Like chlorine, for example, these two bodies unite with hydrogen, producing gaseous compounds respectively called hydrobromic and hydriodic acids, each strikingly analogous to hydrochloric acid, and manifestly constructed upon its type.

Again, there are two elements called respectively Sulphur and

Selenium, presenting many points of analogy with oxygen; and, amongst other things, forming, with hydrogen, gaseous compounds called respectively sulphuretted and selenetted hydrogen, which, though differing in many of their properties from the compound of hydrogen with oxygen, are nevertheless chemical congeners with water, as being unquestionably constructed upon the same type.

So, once more, do the elements Phosphorus and Arsenic present themselves as analogues of nitrogen, when studied in their combinations with hydrogen, forming the gaseous compounds termed respectively phosphoretted and arsenetted hydrogen. At the first glance these hydrogen-compounds present but few points of resemblance with ammonia; but closer investigation reveals a construction so directly comparable, that it is impossible not to recognize, in phosphoretted and arsenetted hydrogen, the reproduced type of ammonia.

Let us endeavour to trace, step by step, the considerations which have led chemists to the admission of these typical relations.

If the view of the analogous construction of hydrochloric, hydrobromic, and hydriodic acids be correct, 2 vols. of hydrobromic and hydriodic acid gases must contain 1 vol. of hydrogen each, united with 1 vol. of bromine-gas and 1 vol. of iodine-gas respectively. Now, this is exactly the result which has been established by experiment; though the method employed to verify the composition of hydrobromic and hydriodic acids differs from that which we found it convenient to adopt in the case of hydrochloric acid.

The circumstances necessitating this change of method may be easily explained. At ordinary atmospheric temperatures chlorine is a gas; and this body consequently lends itself readily to the volumetric determinations which we adopted in its study. But it is far otherwise with bromine and iodine. At ordinary atmospheric temperature the former body is a liquid, the latter a solid body. Hence it follows that the volumetric determination of the quantity of bromine-gas and iodine-gas contained in hydrobromic and hydriodic acids respectively would be attended with considerable difficulty. For this reason it is found preferable, in studying the composition of these substances, to replace the volumetric by the ponderal method of measurement. Indeed, the analogy of hydrobromic and hydriodic acids with hydrochloric acid may be quite as satisfactorily determined by means of the balance, as by direct volumetric analysis.

It is here that we avail ourselves for the first time of this indirect method of ascertaining volumetric relations. As we proceed, however, we shall see that in most instances the balance is our safest and readiest guide to accurate volumetric determinations; while in many cases it is the only means available for use in such inquiries. Hence the method of ascertaining volume-ratios by ponderal analysis deserves, even now, our attentive consideration. It would, however, lead us too far from our immediate subject were we here to enter upon an inquiry into the nature of ponderal analysis, and the means, sometimes simple, oftener very complex and refined, by which the chemist eliminates and weighs in succession the several constituents of a compound body. Nor can we make this method of investigation the subject of rapid experimental demonstrations, such as are alone suited for our purpose during the brief periods of our assembling here to pursue our inquiries together, and such as we found ourselves able to employ in our study of the volumetric method. Ponderal analysis proceeds for the most part by operations too numerous, too protracted, and too nice for performance, in the short time, and with the restricted appliances, at the lecturer's disposal. We must, therefore, reserve for future occasions the experimental study of the methods employed in ponderal analysis; contenting ourselves for the present with its results, and selecting from these such as bear directly upon the subject we have in hand.

Our present inquiry relates to the three compounds, hydrochloric, hydrobromic, and hydriodic acids; and the ponderal analysis of these three compounds has elicited the following results:—

In hydrochloric acid	1 part of hydrogen is combined with	35·5 parts of chlorine.
In hydrobromic acid		80 „ bromine.
In hydriodic acid		127 „ iodine.

In what manner can these ponderal results be made available to display the volumetric construction of hydrobromic and hydriodic acids in comparison with that of hydrochloric acid?

We have seen that

1 part of hydrogen + 35·5 parts of chlorine = 36·5 parts of hydrochloric acid.

And we have also seen that the 36·5 parts of hydrochloric acid thus obtained occupy exactly double the space filled by the unit-volume of hydrogen. The value 36·5 accordingly represented for us the double volume-weight of hydrochloric acid; of which, consequently, the single volume-weight, *i.e.*, the volume-weight relatively to hydrogen, or the sp. gr., is $\frac{36 \cdot 5}{2} = 18 \cdot 25$.

Supposing hydrobromic and hydriodic acids to be similarly constructed, we must have—

1 part of hydrogen + 80 parts of bromine = 81 parts of hydrobromic acid = 2 vols.; whence the volume-weight of hydrobromic acid $= \frac{81}{2} = 40 \cdot 5$; and

1 part of hydrogen + 127 parts of iodine = 128 parts of hydriodic acid = 2 vols.; whence the volume-weight of hydriodic acid $= \frac{128}{2} = 64$.

The experimental determination of the respective volume-weights has justified these previsions.

We remember, lastly, that the weight of chlorine (35·5) which combines with 1 volume-weight of hydrogen, to form 2 volumes of hydrochloric acid, represents the volume-weight of chlorine. Assuming the hydrobromic and hydriodic acids to have an analogous constitution, it is obvious that the weight of bromine (80) and the weight of iodine (127) respectively united with 1 volume-weight of hydrogen to form 2 volumes of each acid, must represent the respective volume-weights of bromine and iodine in the state of gas; or, briefly, their gas-volume-weights, often termed their *vapour-densities*. The comparison of these elements as gases must of course be conducted at temperatures equal or superior to that at which the least volatile of them assumes the gaseous form; and experiment has proved that, at equal temperatures, bromine gas is 80 times, and iodine gas 127 times, heavier than hydrogen gas.

These comparisons can leave no doubt that hydrobromic and hydriodic acids are respectively constructed upon the type of hydrochloric acid; and, if we represent the respective volume-weights of bromine gas and iodine gas by Br = 80 and I = 127, our information regarding the three analogously-constructed compounds may be thus symbolically epitomized :—

Composition of hydrochloric acid.

H	+	Cl	=	HCl

Composition of hydrobromic acid.

H	+	Br	=	HBr

Composition of hydriodic acid.

H	+	I	=	HI

Or, replacing the diagrammatic symbols, as before, by succinct formulæ, we have—

For hydrochloric acid H + Cl. = HCl.

For hydrobromic acid H + Br = HBr.

For hydriodic acid H + I = HI.

On comparing the composition of sulphuretted and selenetted hydrogen with that of water, we arrive at analogous results, capable of similar diagrammatic expression.

Two volumes of water-gas, as we have seen, contain 2 vols. of hydrogen united with 1 vol. of oxygen. If sulphuretted and selenetted hydrogen be constructed on the same type, we must find, in 2 vols. of each of these compounds, 2 vols. of hydrogen united with 1 vol. of sulphur-gas and 1 vol. of selenium-gas respectively. Experiment has verified this anticipation. But, as both sulphur and selenium are solids at the common temperature, becoming gases only under the influence of powerful heat, chemists have preferred ponderal to volumetric analysis for the investigation of these cases also; and the balance has elicited the following results:—

In water	2 parts of hydrogen are combined with	16 parts of oxygen.
In sulphuretted hydrogen		32 „ sulphur.
In selenetted hydrogen		79·4 „ selenium.

From these results the gas-volume-weights of the three compounds are readily computed as follows:—

For water-gas—

2 parts of hydrogen + 16 parts of oxygen = 18 parts of water = 2 vols.; whence the volume-weight of water-gas = $\frac{18}{2} = 9$.

For sulphuretted hydrogen—

2 parts of hydrogen + 32 parts of sulphur = 34 parts of sulphuretted hydrogen = 2 vols.; whence the volume-weight of sulphuretted hydrogen $= \frac{34}{2} = 17$.

For selenetted hydrogen—

2 parts of hydrogen + 79 parts of selenium = 81 parts of selenetted hydrogen = 2 vols.; whence the volume-weight of selenetted hydrogen $= \frac{81}{2} = 40{\cdot}5$.

The two last results, calculated on the assumption that sulphuretted and selenetted hydrogen are constructed on the type of water, coincide exactly with the results obtained by experiment.

But our previous inquiries (comp. p. 50) have also taught us that the weight of oxygen (16), which is combined, in water-gas, with 2 parts (*i.e.*, with the weight of 2 vols.) of hydrogen, represents the volume-weight of oxygen. If sulphuretted hydrogen and selenetted hydrogen be analogously constituted, we are justified in expecting that the weights of sulphur (32), and of selenium (79), which enter into combination with 2 parts of hydrogen, express also the respective volume-weights of the sulphur and selenium gases. Quite recently the volume-weights of sulphur and selenium gases have been accurately determined by experiment; and the results show that, at temperatures at which these elements assume the perfectly gaseous condition, sulphur-gas is 32 times, and selenium-gas 79 times, heavier than hydrogen.

The analogy of the construction of water, of sulphuretted, and of selenetted hydrogen is thus most satisfactorily made out; and, if the volume-weights of sulphur-gas and selenium-gas be respectively represented by S = 32, and Se = 79, the volumetric information afforded, in these cases, by ponderal analysis, may be thus diagrammatically symbolized :—

Composition of water-gas.

$$\left.\begin{array}{c}\boxed{H}\\\boxed{H}\end{array}\right\} + \boxed{O} = \boxed{H_2O}$$

Composition of sulphuretted hydrogen gas.

$$\left.\begin{array}{c}\boxed{H}\\\boxed{H}\end{array}\right\} + \boxed{S} = \boxed{H_2S}$$

Composition of selenetted hydrogen gas.

H, H	+	Se	=	H_2Se

Dropping, as before, our diagrammatic squares, we obtain the following concise expressions :—

For water-gas	$2H + O = H_2O$.
For sulphuretted hydrogen	$2H + S = H_2S$.
For selenetted hydrogen	$2H + Se = H_2Se$.

It now only remains for us to ascertain, in the last place, whether the typical construction of ammonia be really reproduced, in conformity with our anticipation, in phosphoretted and arsenetted hydrogen; but this inquiry, together with some general remarks on chemical symbolization, must be reserved for a separate lecture.

LECTURE V.

Chemical symbols (*continued*)—phosphoretted and arsenetted hydrogen—their construction upon the type of ammonia—ponderal analysis and volume-weights of these compounds—exceptional volume-weights of phosphorus and arsenic gases—combining weights of phosphorus and arsenic—general remarks on chemical symbolization—chemical formulæ as instruments of classification—representation of chemical processes in equations—translation of formulæ into concrete weights and volumes—ponderal analysis of sodic and potassic chlorides, oxides, and nitrides—determination of the combining weights of sodium and potassium.

Having considered the structure of the compounds included in two of our typical groups, those, namely, which are formed in the mould of hydrochloric acid and water, we have now lastly to examine, in like manner, our third typical group, that, to wit, at the head of which we have placed ammonia; our object being to ascertain whether the structural type of this body is really reproduced, as we conceive it to be, in phosphoretted and arsenetted hydrogen.

For this purpose we must bear in mind that, while 2 vols. of hydrochloric acid were found to contain 1 vol. of hydrogen combined with 1 vol. of chlorine; and while 2 vols. of water-gas were found to contain 2 vols. of hydrogen combined with 1 vol. of oxygen; we established that 2 vols. of ammonia contain 3 vols. of hydrogen united with 1 vol. of nitrogen. The question now before us is, whether phosphoretted and arsenetted hydrogen tally precisely with this last-mentioned type?

Here, again, chemists have had recourse to ponderal analysis for the desired information; and the composition of ammonia, phosphoretted and arsenetted hydrogen, as respectively determined by aid of the balance, is as follows :—

In ammonia	3 parts of hydrogen are combined with	14 parts of nitrogen.
In phosphoretted hydrogen		31 „ phosphorus.
In arsenetted hydrogen		75 „ arsenic.

We know, from our previous experiments, that, in the case of ammonia, these weight-results correspond to the relative volume-weight, or specific gravity, subjoined :—

3 parts of hydogren + 14 parts of nitrogen = 17 parts of ammonia = 2 vols.; whence the volume-weight or sp. gr. of ammonia $= \frac{17}{2} = 8{\cdot}5$.

If, therefore, the volumetric structure of phosphoretted and arsenetted hydrogen be identical with that of ammonia, we ought to obtain, as their relative volume-weights, the following respective values :—

For phosphoretted hydrogen :—

3 parts of hydrogen + 31 parts of phosphorus = 34 parts of phosphoretted hydrogen = 2 vols.; whence the volume-weight of phosphoretted hydrogen $= \frac{34}{2} = 17$.

For arsenetted hydrogen :—

3 parts of hydrogen + 75 parts of arsenic = 78 parts of arsenetted hydrogen = 2 vols.; whence the volume-weignt of arsenetted hydrogen $= \frac{78}{2} = 39$.

These are, in very truth, the volume-weights of phosphoretted and arsenetted hydrogen, as furnished by experiment.

So far, therefore, as our inquiry bears upon the three compounds in question, *when already formed*, and not upon their elements *while as yet uncombined*, experiment seems to justify us in affirming the entire identity of the structural type exemplified in these three cases.

But when, from the compounds formed, we turn to consider the volumes of the elements which take part in their formation, we discover a most remarkable and curious discrepancy—the first of its kind that our studies have brought under our notice. To the nature of this discrepancy we must now pay particular attention.

The weight of nitrogen (14) which combines with 3 parts by weight of hydrogen to form ammonia, is, as we have seen, the volume-weight of nitrogen. In other words, the *combining weight* of nitrogen coincides with its *volume-weight.* Assuming the chemical construction of phosphoretted and arsenetted hydrogen to be strictly analogous to that of ammonia, we should expect that the weights of phosphorus (31) and of arsenic (75) united with 3 parts of hydrogen, in phosphoretted and arsenetted hydrogen, would represent the volume-weights of phosphorus and arsenic respectively. In other words, we should expect, in their case, the same coincidence of the *combining weight* with the *volume-weight* as obtains in the case of nitrogen.

Here, however, we meet with the first exceptions to a rule hitherto unbroken.

The volume-weights of phosphorus and arsenic are not represented by the figures (31 and 75), which express their respective combining weights; though, as we shall presently find, the two values stand in a very simple ratio to each other. Upon experimentally comparing the weights of equal volumes of hydrogen, phosphorus, and arsenic, at the temperature at which these latter bodies (usually solid) become gaseous, we find that phosphorus gas is not 31 times, nor arsenic gas 75 times heavier than hydrogen, but that each of these figures has to be *doubled* to bring it into conformity with fact. In other words, the volume-weight of phosphorus gas is not, like its combining weight, 31, but $31 \times 2 = 62$; and the volume-weight of arsenic gas is not, like its combining weight, 75, but $75 \times 2 = 150$.

This striking and singular deviation from a coincidence hitherto constantly observed, stands before us at present unexplained. We are unable to suggest any end likely to be served by these exceptional volumetric relations of phosphorus and arsenic gases, and by the *pro tanto* deviation from the ammonia type thus occasioned in phosphoretted and arsenetted hydrogen. But though the purpose of this difference is as yet unknown to us, its nature and limits are most clearly made out.

Each of the three compounds under consideration contains, in two volumes, three volumes of hydrogen; each also has, on experiment, furnished a volume-weight, corresponding strictly with that calculated from its ponderal analysis; but the quantities of nitrogen, phosphorus, and arsenic combined, in the three compounds, with three parts of hydrogen, are the weights of *unequal* volumes of the three elements: the weight of the nitrogen (14) representing *one volume* of nitrogen, while the weights of the phosphorus (31) and of the arsenic (75) represent only *half a volume* of phosphorus and arsenic gases respectively. These preponderant analogies, and this partial discrepancy, observed in the construction of the three compounds, are conspicuously displayed in the following diagrams, which, it will be observed, are of two kinds; or, rather, which may be said to depict the facts from two points of view.

In the first series the *volume-weights* of phosphorus and arsenic, which, as we have just seen, are exactly double their *combining weights*, are taken as the starting-point of the symbolic expression; and the depicted volumes of phosphorus and arsenic are thus brought into conformity with the unit-volume of hydrogen —our standard or normal volume. This involves, of course, the duplication, as well of the three volumes of hydrogen taking part in the combination, as of the normal product-volume (two unit-volumes) of the compound in each case generated. It also necessitates the adoption of two literal symbols in each case, one to represent the *volume-weight*, the other to denote the *combining weight* of the body in question. For phosphorus we may conveniently employ, to denote these two weights respectively, the symbols Pho = 62, and P = 31; for arsenic, the symbols Ars = 150, and As = 75. The diagrammatic representations, thus brought into conformity with fact, assume the following appearance relatively to ammonia, which is prefixed, as a standard for comparison. Taking first the *volume-weights* as the bases of the symbolic representations, we have :—

Composition of ammonia.

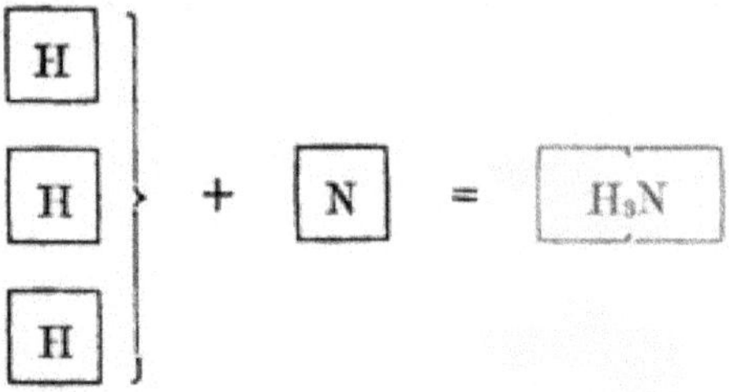

$$\left.\begin{array}{c}\boxed{H}\\ \boxed{H}\\ \boxed{H}\end{array}\right\} + \boxed{N} = \boxed{H_3N}$$

Composition of phosphoretted hydrogen.

$$\left.\begin{array}{cc}\boxed{H} & \boxed{H}\\ \boxed{H} & \boxed{H}\\ \boxed{H} & \boxed{H}\end{array}\right\} + \boxed{Pho} = \boxed{H_6Pho}$$

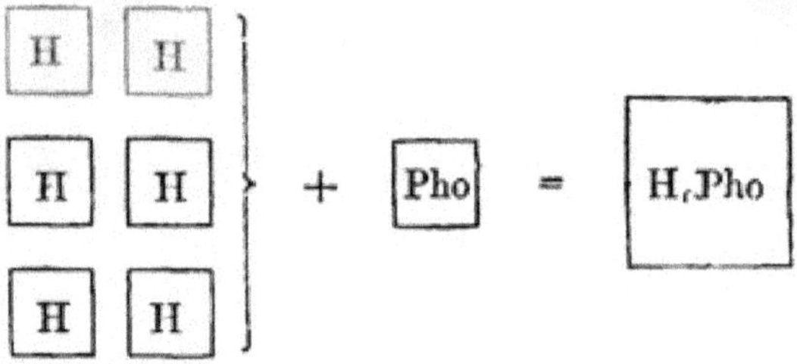

Composition of arsenetted hydrogen.

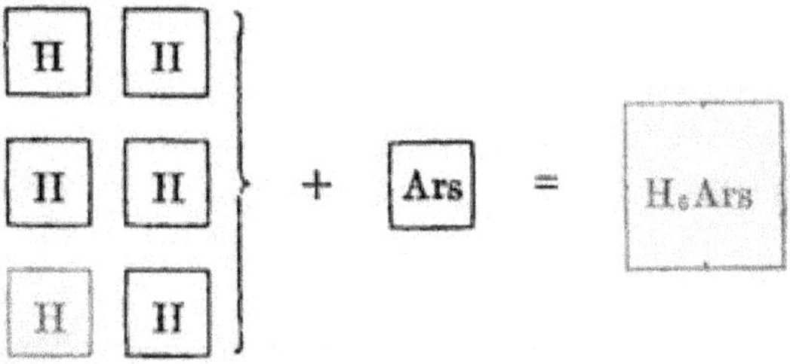

$$\left.\begin{array}{cc}\boxed{H} & \boxed{H}\\ \boxed{H} & \boxed{H}\\ \boxed{H} & \boxed{H}\end{array}\right\} + \boxed{Ars} = \boxed{H_6Ars}$$

Dropping the squares, as before, we obtain the following formulæ:—

For ammonia	$3H + N$	$= H_3N.$
For phosphoretted hydrogen	$6H + Pho.$	$= H_6Pho.$
For arsenetted hydrogen	$6H + Ars.$	$= H_6Ars.$

In the second series of diagrams an opposite starting-point is adopted.

Composition of ammonia.

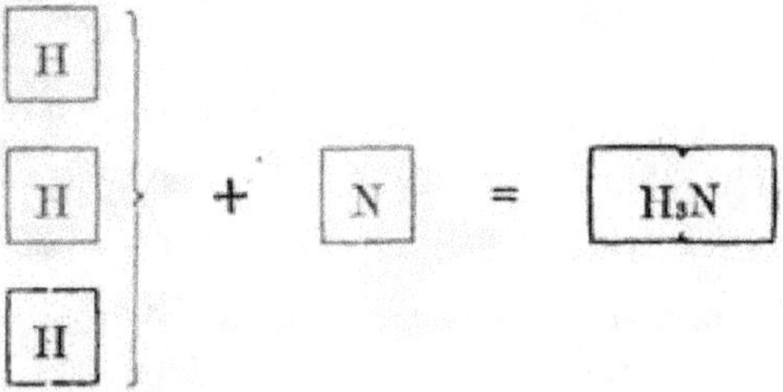

H, H, H	+	N	=	H_3N

Composition of phosphoretted hydrogen.

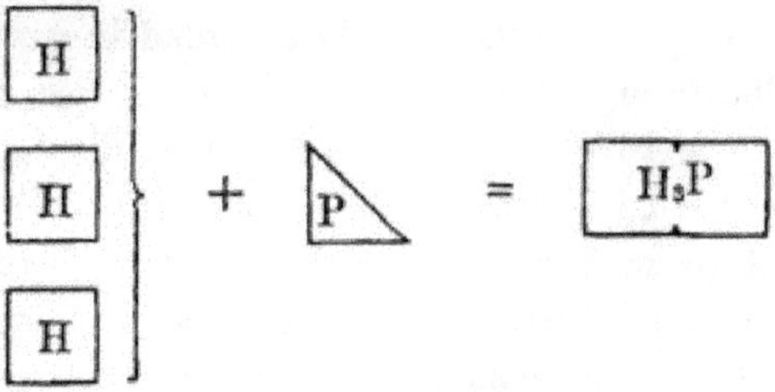

H, H, H	+	P	=	H_3P

Composition of arsenetted hydrogen.

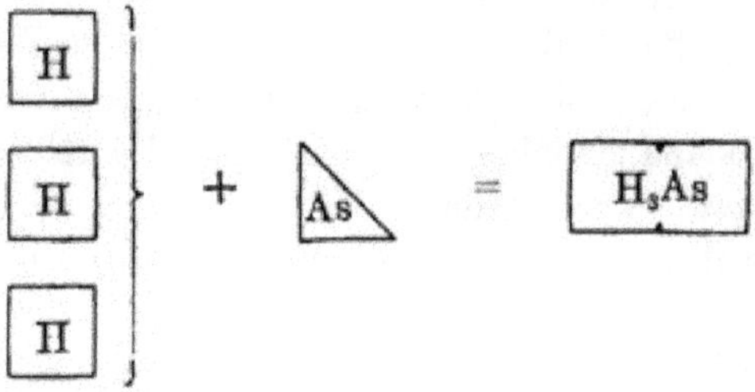

H, H, H	+	As	=	H_3As

It is not the *volume-weights,* but the *combining weights* of phosphorus and arsenic that are here taken as the bases of the respective symbolic representations. Consequently, the volumes of phosphorus and arsenic gases are depicted as of only half the size of the normal unit-volume. This form of diagram obviates the necessity of duplicating the hydrogen volumes and the product-volumes, and gives a series of simpler symbols, in which (as above mentioned) P stands as = 31, and As as = 75; their semi-volumetric relation being represented by a special symbol having only half the area of the square.

Omitting the squares in this, as in the previous case, the formulæ become :—

For ammonia $3H + N = H_3N$.

For phosphoretted hydrogen $3H + P = H_3P$.

For arsenetted hydrogen $3H + As = H_3As$.

It will, of course, be clearly understood that these two modes of symbolization may be employed indifferently; and that neither of them gives us any further insight into the chemical structure of these two compounds, than that which is afforded by the experimental results equally recorded in both forms of diagram. The form exemplified in the first series may be described as the *bi-ponderal* form; while that displayed in the second series may be designated the *semi-volumetric.* The latter form is much to be preferred, as retaining unaltered, in each case, the three unit-volumes of hydrogen, and the normal double-volume of the resultant compound; while the changed appearance of the middle term in the formulæ of phosphoretted and arsenetted hydrogen fixes attention, at the first glance, on the real and only point of difference between these formulæ and the normal ammonia-formula, viz., the semi-volumetric character of phosphorus and arsenic gases, as compared with nitrogen.

We shall have to return on future occasions to the exceptional characters of these phosphorus and arsenic compounds; for the present we commit to memory the experience that, while the symbols H, Cl, Br, I, O, S, Se, and N, respectively express both the *combining weights* and the *volume-weights* of hydrogen, chlorine, bromine-gas, iodine-gas, oxygen, sulphur-gas, selenium-gas, and nitrogen, the symbols P and As express the *combining*, but not the *volume*-weights of phosphorus and arsenic gases; the latter being double the former weight in the case of each of these elements.

Though the complete unity of the symbolic notation is disturbed by these two exceptional cases, the practical value and importance of this admirable language—which may be well termed the Algebra of chemistry—remains entirely unimpaired.

By replacing, for phosphorus and arsenic, the square which expresses a full volume, by the triangle resulting from the diagonal bisection of that square, the semi-volumetric character of these exceptional bodies is appropriately symbolized (as in the above diagram), and all the other symbolic expressions fall under a universal rule.

Of the value of such symbolic formulæ we have already had many examples; and if they possessed no other capability than that, already known to us, of tersely embodying the elementary structure of particular compounds, they would deserve to rank among the chemist's most powerful instruments of *research*.

Instructive, however, as individual formulæ are—shortly as they sketch, so to speak, the portrait of each compound, and vividly as they impress its salient features on the mind—their value becomes still more conspicuous when they are studied in contrast with each other, and employed as instruments of *classification*. By comparing the formula of any given compound with each of the typical formulæ in succession, the group to which the compound belongs is speedily revealed; and from the known chemical characters of the group, the probable properties of a newly-introduced member may be forecast, so as to shape out an appropriate course for research, and not seldom to anticipate its results.

But it is, perhaps, after all, as a *language*—as an instrument of *record*, rapidly written, rapidly read, and capable of presenting collectively to the mind extensive concatenations of fact that words could only convey piecemeal—it is probably in this, which may be termed its *short-hand* application, that the value of chemical formulæ is, in the practical cultivation of the science, most conspicuously manifested.

Nor is it for practical purposes only that chemical formulæ are thus valuable. In the hands of the chemical philosopher they become the expressions, at once concise and comprehensive, of chemical laws—the abstract representations of definite proportionalities—which are transformed into statements of actual facts, when the symbols are replaced by the concrete values for which they stand.

For all these purposes chemical symbols may be handled like ordinary algebraic expressions. The symbolic representations of the various elements and compounds may be connected by the ordinary algebraic signs: by the sign of addition, +; of multiplication, ×; of subtraction, −; and of equality, =.

In this manner we may construct Chemical Equations, by aid of which the most complicated chemical reactions may be rapidly traced out through all their conceivable permutations; each possible chemical change thus set forth suggesting, perhaps, an experiment. On the other hand, the worthlessness of particular schemes of research may be often perceived beforehand, or the fallacies of some too hasty generalization detected without difficulty, by merely subjecting the anticipated reactions to the searching test of a symbolic investigation.

The young chemist cannot, indeed, too early, nor too earnestly, study the language of chemical formulæ. Long before he is far enough advanced in his studies to appreciate their higher philosophical uses, he will find them invaluable aids of his progress, if only to test the clearness and precision of his reasonings, and the accuracy of his experimental results. Results may, indeed, happen to be susceptible of expression in logical symbolic equations, without, on that account, deserving confidence as of necessity accurate; but those which cannot be brought into coincidence with the appropriate formulæ may, with perfect safety, be set aside as inexact. It should, therefore, be the student's object, from the outset of his career, to attain fluency both in reading and writing chemical symbols and formulæ; so that he may be prepared, at a later period, to acquire skill in the art of applying them, for the solution of practical and theoretical problems.

As at once an introduction to this study, and a summary of the reactions which we have already passed in review, we will proceed to record these, more fully and precisely than before, by availing ourselves of the language of chemical formulæ.

The student doubtless bears in mind how useful we found the strong attraction of chlorine for hydrogen, in revealing to us the composition of water and ammonia; and how readily, by means of this powerful agency, we were enabled to withdraw hydrogen from both those compounds, so as to liberate therefrom the oxygen and nitrogen they respectively contain. The formation of hydrochloric acid on the one hand, and the liberation of oxygen and nitrogen respectively on the other, place beyond doubt the *qualitative* composition of the two compounds thus examined.

We must go further than this, however, in order to arrive at a symbolic statement of these changes. We must obtain a *quantitative* as well as a *qualitative* knowledge of the reactions which take place in these experiments. These investigations have been made. Chemists have determined with precision the quantity of chlorine required for the decomposition of a given amount of water and ammonia; the quantity of oxygen and nitrogen which are evolved in each case respectively; and the quantity of hydrochloric acid produced in each case. These valuable results may be concisely embodied in the following simple equations:—

1. *Decomposition of water by chlorine, and products of the reaction:*—

$$\left.\begin{matrix}H\\H\end{matrix}\right\} O + Cl + Cl = HCl + HCl + O;$$

or, more simply,

$$H_2O + 2\,Cl = 2HCl + O.$$

This equation not only illustrates the *qualitative* results of the experiment, but gives a comprehensive synoptical view of the *quantitative* reaction which takes place, and of the weights of the final products. To read this equation, we have only to attach to each letter the name, weight, and volume, for which it stands, when the proportions these bear to each other are at once made manifest. Replacing the symbols by their ponderal values—H by 1, O by 16, and Cl by 35·5, the expression becomes:—

$2 + 16 = 18$ parts of water, require for their decomposition $2 \times 35{\cdot}5 = 71$ parts of chlorine.

While the products are :—

$2 \times (1 + 35{\cdot}5) = 73$ parts of hydrochloric acid, and 16 parts of oxygen.

Reading this equation, again, for the volume-ratios which it also expresses, we learn from it that two volumes of water-gas (the condensed product, as we already know, of the three volumes H + H + O) require for their decomposition two volumes of chlorine; the products formed being $2 \times 2 = 4$ volumes of hydrochloric acid, and 1 volume of free oxygen.

2. *Decomposition of ammonia by chlorine, and products of the reaction.*

This decomposition, like the former one, may be expressed in two ways; we may either write :—

$$\left.\begin{matrix} H \\ H \\ H \end{matrix}\right\} N + Cl + Cl + Cl = HCl + HCl + HCl + N;$$

or, more simply :—

$$H_3N + 3Cl = 3HCl + N;$$

the interpretation of both expressions affording us precisely similar information.

Replacing the symbols by their numerical values, as before, we obtain the following weight and volume ratios :—

$$\underbrace{17 \text{ parts ammonia}}_{2 \text{ vols.}} + \underbrace{3 \times 35{\cdot}5 \text{ parts chlorine}}_{3 \text{ vols.}}$$

$$= \underbrace{3 \times (1 + 35{\cdot}5) \text{ parts hydrochloric acid}}_{6 \text{ vols.}} + \underbrace{14 \text{ parts nitrogen}}_{1 \text{ vol.}}$$

All this information, be it observed, is symbolized in the preceding equation by eleven letters and figures, distributed into four groups, joined by three relational signs. It is difficult to conceive a form of expression more concise and encyclopædic.

In the processes just reviewed we have had exclusively to deal with elements which, at the common temperature, are gases; and the volume-weights of which are, therefore, easy of determination. This gaseous form does not, however, characterize all the elementary bodies, and hence arises a new and serious obstacle in our path, when we endeavour to extend the application of our newly-acquired language to solids. We need not go beyond our present limited range of experience for examples of this difficulty. It meets us even in the three simple experiments which we selected to inaugurate our chemical studies.

It will be remembered that, in order to separate the hydrogen from hydrochloric acid, from water, and from ammonia, we submitted these compounds successively to the action of two metallic elements, sodium and potassium, both of which are solid bodies, not gases. The sum total of the information furnished by these experiments, under the conditions stated, was the fact of the evolution of hydrogen. The true nature of these reactions, the changes undergone by the sodium and potassium employed, the weight-ratios in which the substances present act upon one another, did not in those experiments press for immediate consideration. These particulars come now in their turn to be investigated, and embodied in appropriate symbolic expressions; and, in setting ourselves to this task, we shall acquire our first experience of the difficulty referred to above, as also of the means by which it is to be overcome.

And first, as regards the *qualitative* character of these reactions. With the experience acquired in studying the action of chlorine upon water and ammonia, we can no longer have any doubt in this respect. Sodium and potassium liberate the hydrogen from hydrochloric acid, water, and ammonia, by withdrawing therefrom the elements chlorine, oxygen, and nitrogen, respectively united therewith. The action of sodium upon hydrochloric acid gives rise to the formation of a solid chlorine compound of sodium, which we call *chloride of sodium*. The action of sodium upon water, under appropriate circumstances, produces an oxygen compound of sodium, also solid, termed *oxide of sodium*. Lastly, the action of sodium, under favourable circumstances, upon ammonia, gives rise to the formation of a third solid compound of sodium; that, namely, which it forms with nitrogen, and which we may therefore call *nitride of sodium.*

But we have now, as before, to represent these processes in equations, capable of disclosing the *quantitative* conditions of the reactions. And here arises our difficulty. At this point we meet with the obstacle opposed to our progress by the solidity, at common temperatures, of one of the elements with which we have to work. For, to accomplish our purpose—to perfect our knowledge of these reactions, and to embody our completed knowledge in volumetric equations—we evidently require to know the gas-volume-weight of sodium; in other words, the specific gravity of sodium-gas. Unfortunately, sodium can only be volatilized at very high temperatures and under conditions which make it extremely difficult to obtain the gas pure, and to ascertain its volume-weight. Sodium-gas has never yet been obtained in a state of perfect freedom from admixture, and its volume-weight is consequently unknown to us.

In the absence of direct means of ascertaining the volume-weight of this metal, Chemists have had recourse to an indirect mode of research. They have striven to ascertain its combining weight relatively to gaseous bodies; and, for this purpose, they have sought to obtain a gaseous combination of sodium with hydrogen. Could this be accomplished, the normal product-volume of the compound gas (our well-remembered double unit-volume) might be measured out and analyzed, and the weight of sodium contained in it would represent the combining weight of this body relatively to our standard, hydrogen. But this resource also has failed. Chemists have hitherto been unable to produce a hydrogen-compound of sodium.

Under these circumstances, nothing has remained but to fall back upon ponderal analysis of the solid compounds of sodium—such as the chloride, the oxide, and the nitride of this body, referred to in our previous experiments. With the value of this expedient we are already acquainted. Ponderal analysis, pure and simple, has already rendered us important service in the study of hydrobromic and hydriodic acid, and of sulphuretted and selenetted hydrogen; while in the investigation of phosphoretted and arsenetted hydrogen we had to rely almost exclusively upon its indications.

Let us now see how far this method will serve our purpose in the cases under consideration.

Weight-analyses of the three compounds generated by the action of sodium upon hydrochloric acid, water, and ammonia, have furnished the following results :—

In chloride of sodium—

35·5 parts of chlorine (the weight of 1 vol.), are combined with 23 parts of sodium.

In oxide of sodium—

16 parts of oxygen (the weight of 1 vol.), are combined with $23 \times 2 = 46$ parts of sodium.

In nitride of sodium—

14 parts of nitrogen (the weight of 1 vol.), are combined with $23 \times 3 = 69$ parts of sodium.

Equal volumes of chlorine, oxygen, and nitrogen, are thus seen to combine with very unequal weights of sodium. The weight of sodium combining with one volume of oxygen is *twice*, the weight combining with one volume of nitrogen is *thrice*, the weight combining with one volume of chlorine. We thus see that, in its behaviour towards chlorine, towards oxygen, and towards nitrogen, sodium exactly resembles hydrogen; for we remember that one volume of chlorine, of oxygen, and of nitrogen, respectively combine with one, two, and three parts by weight of hydrogen.

With these results before us, and bearing in mind the experimental proofs already obtained that the weights of bromine and of iodine respectively combining with one volume of hydrogen actually represent the gas-volume-weights of those elements, are we justified in admitting the weight of sodium combined with one volume of chlorine to be the volume-weight of sodium gas? Further experiment is needed to answer this question. There is, however, no valid objection to our provisional acceptance of this conception; and, in this sense, we may adopt, as the gas-volume weight of sodium, the value 23, representing the weight of sodium which, in chloride of sodium, is united with one volume of chlorine. The symbol of sodium is derived from the Arabic name for soda, *natron*, whence *natrium*, the German name for this metal; the complete expression for this element is accordingly Na = 23.

Thus much being taken for granted, the composition of the three sodium compounds may be represented in formulæ, capable of tersely and clearly setting forth their analogies with the hydrogen compounds whose types of structure they respectively affect. Thus, placing in one column the typical compounds, and in the other their analogues, so as to bring them into comparison, side by side, we obtain the following series:—

TYPES.	ANALOGUES.
Chloride of hydrogen (hydrochloric acid) HCl.	Chloride of sodium (common salt) $NaCl$.
Oxide of hydrogen (water) $\left.\begin{matrix}H\\H\end{matrix}\right\}O = H_2O$	Oxide of sodium (soda) $\left.\begin{matrix}Na\\Na\end{matrix}\right\}O = Na_2O$.
Nitride of hydrogen (ammonia) $\left.\begin{matrix}H\\H\\H\end{matrix}\right\}N = H_3N$.	Nitride of sodium $\left.\begin{matrix}Na\\Na\\Na\end{matrix}\right\}N = Na_3N$.

Whether the combining weight of sodium, represented by the expression Na = 23, coincide with the relative volume-weight of its gas; or whether (as we found in the cases of phosphorus and arsenic) the volume-weight of sodium be double its combining weight, so as to correspond to the expression $23 \times 2 = 46$; or whether, lastly, these two values stand in some less simple ratio to each other; these are questions which future experiments can alone decide.

But, whatever may be the result of such ulterior investigations as to the gas-volume of sodium, we are enabled by our present knowledge of its combining weight to represent, in the following simple equations, the action of this metal upon hydrochloric acid, water, and ammonia, respectively :—

$$HCl + Na = NaCl + H$$

$$H_2O + 2Na = Na_2O + 2H$$

$$H_3N + 3Na = Na_3N + 3H$$

Whilst formerly we only knew sodium to be capable of liberating hydrogen from its compounds with chlorine, oxygen, and nitrogen, we now learn, from the above equations, that in order to disengage one volume of hydrogen (H = 1 by weight) we invariably require 23 parts by weight of sodium (Na); this proportionality still holding good, whether the hydrogen compound submitted to the action of the metal contain one, two, or three volumes of hydrogen. We note this fact in passing; its full importance will become apparent to us a little further on in our inquiry.

It is not, however, by sodium only that hydrochloric acid, water, and ammonia, are decomposed; other metals act on them in a similar manner. If, for example, potassium be substituted for sodium in the foregoing experiments, hydrogen is equally disengaged; its evolution being of course attended by the formation of potassic instead of sodic chloride, oxide, and nitride. Ponderal analysis of these compounds, coupled with a series of considerations analogous to those we entered into in the case of sodium, have led chemists to fix the combining weight of potassium at 39 = K (the initial of *kalium*, a term of Arabic origin, retained in the German as the name of the metal). The three formulæ—

$$KCl, \quad \left.\begin{matrix}K\\K\end{matrix}\right\}O, \quad \text{and} \quad \left.\begin{matrix}K\\K\\K\end{matrix}\right\}N,$$

represent these three potassium compounds respectively; while the equations—

$$HCl + K = KCl + H$$
$$H_2O + 2K = K_2O + 2H$$
$$H_3N + 3K = K_3N + 3H$$

illustrate the formation of these potassium compounds from the corresponding hydrogen-compounds. These expressions may be cited as further evidence of the terseness, lucidity, and precision, introduced by the use of the language of symbols into the statement and investigation of chemical transformations.

It is not, however, pretended that this system of chemical notation has arrived at the perfect symmetry, and logical consistency throughout, which we may hope it will attain with the progress of chemical knowledge. At present, some of the relations which it is employed to express are assumptions based only on highly-probable analogies. We have already seen, for example, that, while the four symbols H, Cl, O, and N, represent the volume-weights of elements which are gaseous at common temperatures and pressures, the four symbols Br, I, S, and Se, refer to four elements which are not gases under ordinary barometric and thermometric conditions, and whose weights, under those conditions, can only be arrived at by deduction from their observed volume-weights at the much higher temperatures which do actually convert them into gases. Again, while all the eight symbols above mentioned express the weights of equal volumes of the several elements referred to, the two symbols P and As express only the half-volume-weights of the bodies which they are employed to denote. So, once more, the symbols Na and K can be taken to represent the unit-volumes of the alkaline metals they denote, only by pure assumption; seeing that experiment has hitherto failed to determine the respective volume-weights of those metals, when converted into gas by intense heat.

In these instances, therefore, as well as in the cases of all bodies (unfortunately, a large majority of the known elements) which cannot be volatilized, or rendered gaseous, even by the intensest temperatures as yet at our command, we are compelled to rely exclusively upon ponderal analysis to determine their combining proportions, and from these to infer, by aid of analogy, their probable volume-weights, assuming them to be susceptible of volatilization. In all these instances we are liable to be misled by analogy; to assume, for example, coincidence of the *combining* weight with the *volume*-weight of the body in question (as observed in the cases of bromine and iodine, of sulphur and selenium); whereas, in truth, there may be, as in the cases of phosphorus and arsenic, no such coincidence, but an exceptional divergence; the combining weight corresponding to the half volume only, not to the full volume of the gas; or even, perhaps, standing in some less simple ratio thereto.

It is our duty to bear in mind that, in this, as in all other respects, and in our own as in all antecedent ages, an imperfect and transitional condition still characterizes scientific progress, and impresses a correspondingly uncertain and provisional character upon our best efforts to interpret and extend it.

It is only, therefore, in this limited sense, and subject to these philosophical reservations, that the assumed coincidence of the volumetric with the ponderal structure of chemical compounds is advanced in those cases which, for the present, lie beyond the range of experimental verification.

LECTURE VI.

Fourth term in the series of typical hydrogen-compounds—marsh-gas, or light carbonetted hydrogen—reasons for its separate consideration—its occurrence in marshes—in coal-mines—in coal-gas—its preparation—its distinctive characters—its qualitative analysis—separation therefrom of carbon by chlorine.—Decomposition of marsh-gas by heat—its quantitative analysis—its synthesis not yet accomplished—combining weight of its constituent, carbon—its analogues.—Silicetted hydrogen—probability of the construction thereof upon the type of marsh-gas.

In the series of typical hydrogen compounds which we have now studied, we have found one volume each of chlorine, oxygen, and nitrogen, united respectively with one, two, and three volumes of hydrogen; the condensation increasing in direct ratio with the increasing proportion of hydrogen, so that, in each case, two volumes only of the resultant compound are produced.

This typical series does not end here, however. It comprises a fourth member—a compound containing, in two volumes, four volumes of hydrogen condensed; the other constituent being the body familiar to all as *carbon*. This compound is the light, inflammable gas, too well known to coal-miners as *Fire-damp;* and also termed, on account of its frequent emanation from boggy ground, *Marsh-gas.*

The description of this fourth typical hydrogen-compound might have been included with that of the rest of the series, but we have preferred to reserve it as a matter for separate study, chiefly on account of this signal difference; that whereas in all the other members of the series *both* constituents of the compound are gaseous, in this fourth one the hydrogen is united with a body which is not only solid at common temperatures, but incapable of volatilization by any, the intensest means at our present command. Hence, while the first three compounds, studied collectively, apart from the fourth, illustrated, with an admirable symmetry, and in an unbroken ascending scale, the laws of combination and progressive condensation, by volume as well as by weight, the fourth exemplifies volume combination and condensation only so far as its gaseous ingredient is concerned; our *positive* knowledge, as to its solid constituent, extending only to the *weight*-ratio, and any views we may entertain as to the *volume*-ratio of this element being of necessity speculative. It is, indeed, maintained by many chemists, and not without some show of reason, that analogy affords a fair basis for such speculation; but not until carbon shall have been actually volatilized, and its vapour actually weighed, shall we be entitled to rank marsh-gas, with respect to its volumetric constitution, on the same certain footing as experiment already assigns to hydrochloric acid, water, and ammonia.

In describing this as the *chief*, we imply that it is not the *only* difference which justifies us in separating the study of marsh-gas from that of the three other typical compounds. A peculiarity in its chemical deportment also distinguishes marsh-gas from the other members of the typical series, in a manner which will call for our special attention hereafter, but of which it would be premature to say more, at present, than that its effect is to prevent our employing, in the case of marsh-gas, certain modes of investigation, which we have adopted with advantage in studying the other typical compounds.

Thus much premised, let us proceed to make acquaintance with this fourth member of our typical series.

There escapes from the fissures of the great coal-measures a transparent, colourless, inflammable gas, which frequently accumulates in the galleries of ill-ventilated coal-mines; and, when ignited by the miner's candle, through neglect of the protective precautions provided by science, gives rise to the explosions so much dreaded and deplored on account of their life-destroying violence. Many varieties of coal so abound in this gas, that it may be seen rising therefrom in bubbles, when newly-dug fragments are thrown into water. The same gas, as we have already mentioned, is developed in marshy lands, and is often observed bubbling up from stagnant pools and swamps, where vegetal matters are in process of gradual decay. In summer, when decay is most active, the development of this gas is most abundant; and it may be readily collected in a glass cylinder, inverted over, and plunging into, stagnant waters. The gas thus obtained may be at once distinguished from common air by applying a light, when it is found to be inflammable.

It would, of course, be inconvenient to procure this compound from either of its natural sources, even if they furnished it in a state of purity, instead of, as always happens, yielding it in admixture with common air and other gases. The ordinary illuminating gas, distilled from coal by artificial heat, always contains a large percentage of marsh-gas; but here also it is mixed with other gases, scarcely separable from it; so that we are debarred from this source of supply, otherwise so accessible and abundant. Chemists have, however, devised a simple process by which marsh-gas may be readily prepared in any quantity requisite for laboratory purposes, from well-known materials, procurable at relatively moderate cost. In a flask (of glass, or better, of copper or iron), arranged for gas disengagement, strong vinegar is heated with a mixture of lime and the caustic soda of commerce; after a short time a transparent colourless gas is evolved, which is collected over water in the usual way (Fig. 60). We have, for the present, no interest in dwelling on the reaction which, under these circumstances, gives rise to marsh-gas. Suffice it to say that a portion of the carbon contained in the vinegar combines with the hydrogen present, so as to form this gas; which is also frequently called light carburetted or carbonetted hydrogen. For brevity's sake, however, we will, on the present occasion, retain the name of marsh-gas.

From the hydrogen compounds previously examined, marsh-gas is readily distinguishable by its inflammability. Kindled at a taper it burns with a feebly-luminous flame. Independently of this character, however, marsh-gas is strikingly distinguished from hydrochloric acid and ammonia gases, by its possessing neither odour nor action on vegetal colours; which negative characters it shares with water-gas.

FIG. 60.

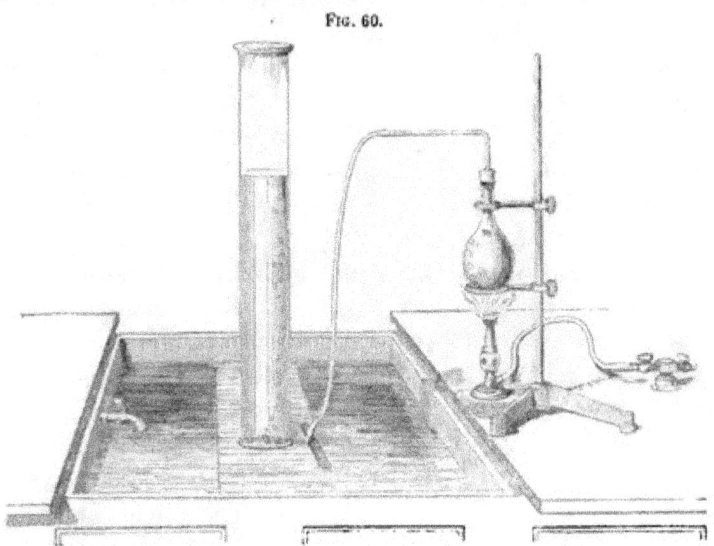

Not less easily may marsh-gas be distinguished from the three elementary gases which we have submitted to examination. Indeed, this readily inflammable gas cannot possibly be confounded either with chlorine, oxygen, or nitrogen, which are all incombustible bodies; the two former being supporters of combustion, but themselves not inflammable, while the last-named neither burns, itself, nor sustains the combustion of other bodies. From chlorine, moreover, marsh-gas differs by its lack of colour, odour, and bleaching property; from oxygen, marsh-gas, like nitrogen, is distinguishable by its total incapacity for supporting the process of combustion. The only elementary gas for which, at the first glance, the hydrogen compound of carbon is at all likely to be mistaken, is hydrogen itself; seeing that inflammability, coupled with the absence of colour, of odour, of bleaching power, and of the property of supporting combustion, characterise both these gases. The difference, however, of hydrogen from carbonetted hydrogen becomes at once obvious if both gases be burned side by side: for, while hydrogen burns with a non-luminous and scarcely-visible flame, the combustion of marsh-gas is attended by the evolution of feeble, but unmistakeable rays of light.

The chemical difference of the two gases may be demonstrated by a simple experiment. We remember that a mixture of hydrogen and chlorine, when ignited, merely gave rise to the formation of hydrochloric acid. A similar experiment performed with a mixture of marsh-gas and chlorine produces, as we shall see, a further and very striking result.

For this purpose, we fill a tall cylinder with warm water, invert it over the pneumatic trough, and pass marsh-gas into it until a little more than one-third of the water is displaced; which done, we fill the two remaining thirds of the cylinder with chlorine, taking care to protect the vessel, during manipulation, from exposure to sunlight. The cylinder thus filled we close, by slipping a glass plate beneath its mouth; then, raising it from the trough, we agitate it to mix the gases thoroughly; and finally, we apply a light to the mixture. Ignition takes place, with production of hydrochloric acid, by the union of the chlorine with the hydrogen present, as in our former experiment with pure hydrogen and chlorine. But the marsh-gas betrays its additional constituent, by a copious separation of carbon as the flame descends into the cylinder; the sides of which become coated with a dense black layer of soot (Fig. 61).

Fig. 61.

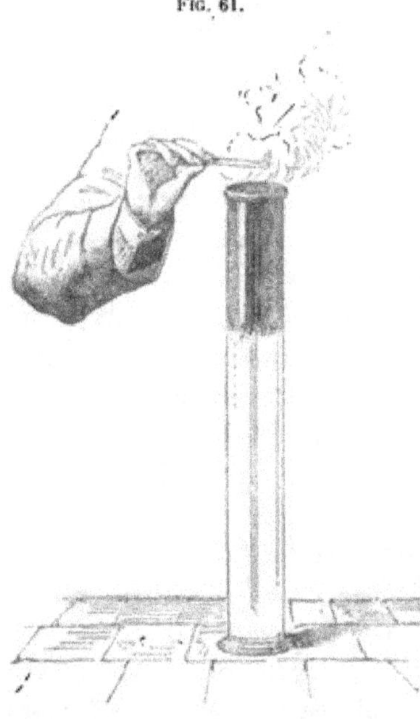

The interpretation of this phenomenon presents no difficulty. Chlorine acts upon marsh-gas as, under appropriate conditions, we have seen it acting upon water and ammonia. Water-gas, under the influence of chlorine, yields its oxygen, and ammonia its nitrogen; in a similar manner, marsh-gas gives off its carbon. The hydrogen withdrawn, in all three cases, combines, in the very act of separation, with chlorine, to form hydrochloric acid gas.

The action of chlorine upon marsh-gas indubitably points to hydrogen and carbon as elementary constituents of this compound; and it therefore remains only to demonstrate, by synthesis, that hydrogen and carbon are its *exclusive* constituents. But here we meet with difficulties very similar to those which compelled us to forego the synthesis of ammonia. No means are known at present by which hydrogen and carbon can be directly combined to form marsh-gas. Nevertheless, in this case, as in that of ammonia, the balance comes to our aid, and enables us to decide the question. The sum of the weights of the two constituents separated from a given quantity of marsh-gas is found to be precisely equal to the weight of the marsh-gas submitted to experiment.

Our next task is to determine the combining weight of the carbon; and, for this purpose, to investigate the ratio in which this body is united with hydrogen in marsh-gas.

Were carbon a gaseous element, we should, of course, have no difficulty in ascertaining its combining weight, by the volumetric method employed in the cases of chlorine, oxygen, and nitrogen. Were carbon even susceptible of volatilization, we might endeavour to determine its volume-weight directly by aid of the balance; though here we should be liable to meet with just such a disparity between the combining and volume weights as exceptionally obtains in the case of phosphorus and arsenic. But, in point of fact, as we have already seen, carbon is a solid which it is not as yet in our power to convert into gas; and therefore we are constrained to fall back, in its case, upon the only *universally* applicable method with which we have thus far become acquainted, for ascertaining, the combining weights of elementary bodies—viz., the determination of the weight thereof contained in the product-volume of that one of its hydrogen-compounds in which it is present in the smallest quantity.

Accordingly, we represent by C the quantity of carbon contained in 2 vols. of marsh-gas; and this we admit as the combining weight of carbon. The task before us resolves itself, therefore, into the exact determination, by means of the balance, of the quantities of hydrogen and carbon respectively

FIG. 62.

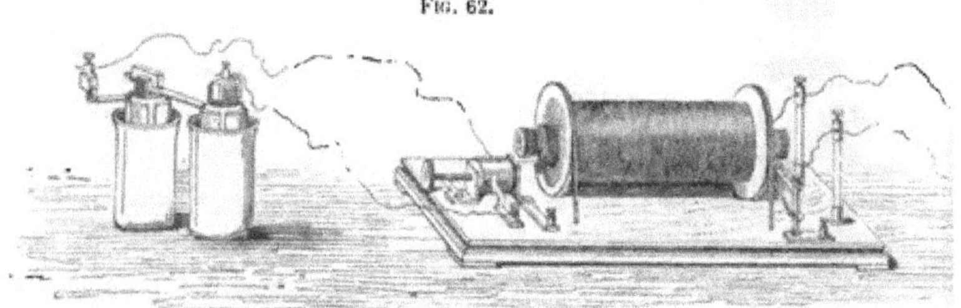

contained in 2 volumes of marsh-gas. The solution of this problem, as we shall hereafter see, is furnished by a single experiment, not less elegant than conclusive. Meanwhile we must be contented to employ a somewhat circuitous and less rigorous method, lying within the scope of our present chemical experience.

For this purpose, our first care is to ascertain the volume-weight of marsh-gas. That this gas is much lighter than atmospheric air is readily seen by simply filling a cylinder with marsh-gas, allowing it to stand uncovered for a few minutes, and then applying a lighted taper. Ignition does not ensue; every trace of the inflammable gas is found to have escaped; the heavier common air having taken its place in the vessel. Careful experiments have proved that marsh-gas is exactly 8 times heavier than hydrogen; hence, taking the unit-volume of hydrogen as 1, two volumes of marsh-gas weigh 16.

How much hydrogen is present in these 2 volumes? This question admits of being answered, in very close approximation, by the method which served us in the case of ammonia, viz., by splitting it up into its constituents under the influence of heat. This experiment we perform, as before, in a U tube, fitted up with spark wires (Fig. 62). As our source of heat, we again employ the spark-current of the induction-coil. No sooner does the spark-current begin to traverse the tube, than the marsh-gas is found to expand; and, after the lapse of a few minutes, a light deposit of carbon is formed in the vicinity of the platinum wires. The decomposition, energetic at the commencement of the experiment, proceeds more slowly as the gas dilates; so that a considerable time is required to bring the operation to a close. If the mercury be allowed to run out from the nipper-tap, till it has become level in the two limbs of the U tube, it is found that the original gas-volume has very nearly doubled. This result having been attained, the continued transmission of the spark-current produces no further increase in the volume of the gas; which is then readily shown, by the usual tests, to have lost the characteristic properties of marsh-gas; to be, in fact, pure hydrogen.

Fig. 62.

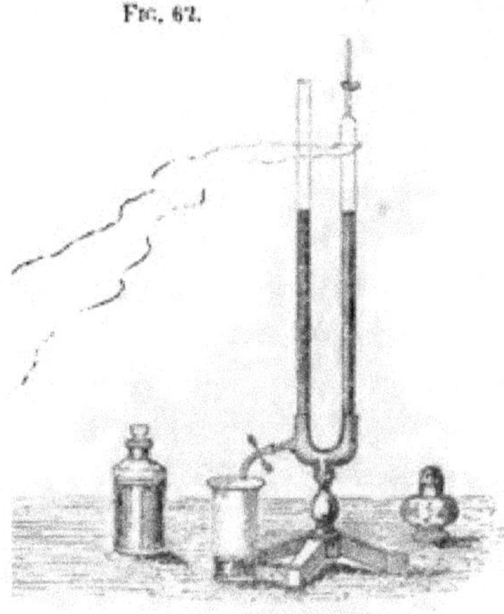

This experiment presents greater difficulty of manipulation than attends the processes we employed for determining the composition of hydrochloric acid, water, and ammonia. The spark-current, when transmitted through marsh-gas, is not unfrequently interrupted by the formation of a conducting bridge of carbon between the platinum points; and this conductor has, of course, to be destroyed, in order that the current may have an interval through which to leap in the form of sparks. In order to break the bridge, we may agitate the mercury in the tube until it reaches the platinum wires. Or, better still, we may endeavour to prevent the formation of the carbon-bridge by reversing from time to time the direction of the current.

With these precautions, the experiment affords very closely approximative, though not perfectly exact, results. Slight error arises from the circumstance that a minute proportion of marsh-gas escapes decomposition by the spark-current; and, accordingly, the volume of hydrogen obtained is never quite double the volume of the marsh-gas submitted to the treatment. Nevertheless, the experiment, even in its present imperfect form, unequivocally points to a result, which hereafter, by other methods, we shall establish beyond question; namely, that marsh-gas contains twice its volume of hydrogen.

Accordingly, two volumes of marsh-gas, the weight of which we have found to be 16, contain 4 volumes (and also, therefore, 4 parts by weight) of hydrogen, combined with 16 − 4 = 12 parts, by weight, of carbon. This quantity (C = 12) is, therefore, the *combining weight* of carbon, *i. e.*, the weight of carbon contained in two volumes of marsh-gas. We are now, therefore, justified in representing the composition of marsh-gas by the diagrammatic equation—

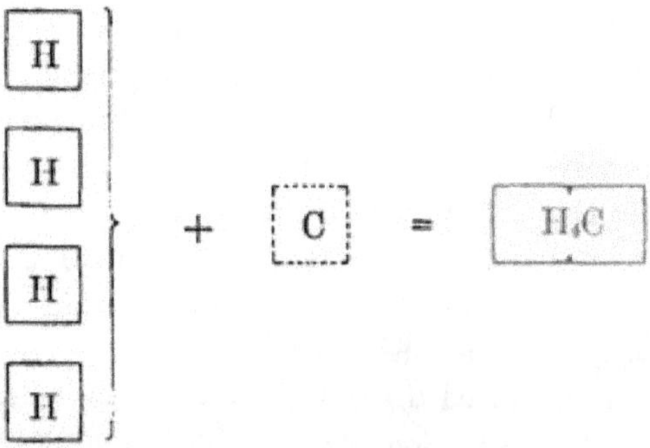

$$\left.\begin{matrix}\boxed{H}\\ \\ \boxed{H}\\ \\ \boxed{H}\\ \\ \boxed{H}\end{matrix}\right\} + \boxed{C} = \boxed{H_4C}$$

Dispensing with the squares, we obtain the short formula—

$$4H + C = H_4C,$$

the second term of which represents the weight of two volumes of this gas, exactly as the expressions HCl, H_2O, and H_3N express the weights of two volumes of hydrochloric acid, water-gas, and ammonia.

It will be observed that, in the diagram, we have surrounded the element C with a square formed in dotted, instead of full lines; an expedient which we shall find convenient on future occasions, as well as in the present case, to signify that the volume represented is hypothetical. The full lines will always be reserved to symbolize volumes placed beyond question by actual experiment.

In marsh-gas we have added a new term to the series of hydrogen-compounds which have so long engaged our attention. Hydrogen does not, so far as we yet know, form with any element whatsoever any compound, whereof the normal product-volume contains more than four volumes of hydrogen united with the combining weight of the other element. Marsh-gas may, therefore, be considered as an example of the hydrogen-compounds richest in hydrogen; exactly as we regard hydrochloric acid as exemplifying the hydrogen-compounds poorest in hydrogen; and water and ammonia as representing hydrogen-compounds of intermediate composition. These relations are clearly displayed in the following series of formulæ:—

Hydrochloric acid ..	$H\,Cl$	=	2 vols.
Water	H_2O	=	2 „
Ammonia	H_3N	=	2 „
Marsh-gas	H_4C	=	2 „

It remains to justify the introduction of marsh-gas into the series of *typical* hydrogen-compounds; in other words, to show that it deserves to be regarded as the model upon which one or more analogous compounds are formed.

In order definitively to raise marsh-gas to this rank, we ought, evidently, to be in possession of some binary hydrogen-compound or compounds, each containing some constituent analogous to carbon, and, moreover, containing it in such relative proportions that, with its minimum combining weight there may be united four volumes of hydrogen; the whole being reduced, by condensation, to one normal product-volume.

Only one such analogue of marsh-gas has hitherto been actually obtained; and even this is but a recently discovered and as yet imperfectly studied compound. It is the compound of hydrogen with silicon; and so far as experiment has yet made out its composition, this body appears to contain—

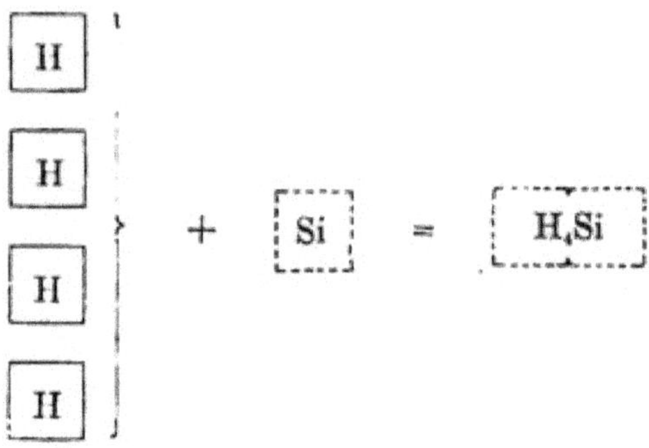

Or, dropping the squares as before,

$$4H + Si = H_4Si;$$

in which Si (= 28·5) represents the combining weight of silicon.

That silicon does really thus combine with hydrogen in the ratio above represented, is rendered all the more probable by the fact that, in forming compounds with other elements, carbon and silicon resemble each other as to the proportions of such elements which they respectively take up. Analogies of a similar kind justify the anticipation that hydrogen-compounds of titanium and tin will be obtained, having the same proportional composition as the hydrogen-compound of silicon, and forming, with it, a normal series of hydrogen-compounds grouped under the type of marsh-gas. These elementary bodies have not yet been volatilized, it is true; but so neither has carbon itself. This circumstance, therefore, will only oblige us to adopt, as their combining weights, so much of each body as the balance may prove to exist in two volumes of the hydrogen-compound it may be found to produce; guarding ourselves against the symbolic affirmation of more than we actually know, by representing, as before, the assumed values by dotted squares.

From the series of phenomena which we have now studied together, we have already begun to deduce some valuable general conceptions. By the experimental examination of a small number of bodies we have learned to contradistinguish *elementary* bodies from *compounds*, and *mechanical mixture* from *chemical combination*. We have also been led, step by step, to recognize the conditions of immutable proportionality under which chemical compounds are formed, and the simple volume-ratios in which the elements associate for their production. The comprehension of these ratios induced us naturally to seek a concise yet exact language for their expression, and this we found in the symbolic system of chemical notation. Referring to experiments, few but conclusive, for the proof of our positions, and the guidance of our progress, we were enabled to examine, by their certain light, results gathered in a much wider sphere than we have yet mastered for ourselves; and thus, by the alternate employment of the inductive and the deductive methods, we have been enabled to construct a general system, in which many particular facts may be embraced.

This is something to have achieved, but much more remains to be done. Let us bear in mind that our notions rest as yet upon the study of a very limited number of bodies, elementary and compound; let us be prepared for considerable extension, and at the same time clearer definition of our views, as our knowledge of the facts to be co-ordinated acquires a wider range.

Meanwhile we have still much to learn without going beyond the five elements, hydrogen, chlorine, oxygen, nitrogen, and carbon, to which we have confined our attention. Thus far we have examined only the typical compounds formed by hydrogen with the four other elements just enumerated; and these typical hydrogen compounds are doubtless of paramount importance as forming the main pillars of our edifice. Nevertheless, we shall have hereafter to make acquaintance with a countless variety of compound bodies, in whose composition hydrogen bears no part; and these are well exemplified in the varied combinations formed amongst each other by chlorine, oxygen, nitrogen, and carbon. The exhaustive study, even of this narrow field, would go far to prepare us for mastering the much wider domain that lies beyond; but for the purposes of the present introduction—necessarily restricted in its scope—we will select from among these manifold subjects one only for immediate consideration—that, namely, which relates to the deportment of nitrogen towards oxygen.

LECTURE VII.

Deportment of nitrogen towards oxygen—nitric acid—hydrated—anhydrous—its composition—its decomposition—by heat—by metals—by tin, yielding hyponitric acid—by silver, yielding nitrous acid—by copper, yielding nitric oxide—by zinc, yielding nitrous oxide—characters of these products—how shown to be chemical compounds, not mechanical mixtures—expansion of the idea of chemical combination—combination of two elements in several proportions—law of multiple proportions—volume and condensation ratios in chemical compounds—ordinary—exceptional.

Having selected as the next subject of our inquiry the deportment of nitrogen towards oxygen gas, we are led at once to the study of one of the most intense and interesting agents existing in the whole range of chemistry—namely, Nitric acid.

Nitric acid is a powerfully-corrosive liquid, long known by the name Aquafortis, and now manufactured on a large scale for industrial purposes. This liquid contains, together with water, a compound of nitrogen and oxygen, which, when entirely separated from water, is called anhydrous nitric acid, or nitric anhydride. This separation is not, however, so easily effected as in the case of hydrochloric acid; the aqueous solution of which, as you remember, only required heating to ebullition for this purpose.

By appropriately selected means, the examination of which we must defer, the dehydration of nitric acid has been nevertheless accomplished.

Anhydrous nitric acid presents itself in the form of white fusible crystals, in which analysis has proved the presence of two volumes of nitrogen combined with five volumes of oxygen, or 28 parts by weight of the former with 80 parts of the latter element. Anhydrous nitric acid is a somewhat instable compound. It is decomposed even by gentle heat, with evolution of copious red fumes, which, when passed through water, are for the most part absorbed, the unabsorbed portion being a colourless gas, which is readily recognised as oxygen.

It is not, however, in its anhydrous condition only that nitric acid is thus easy of decomposition. The ordinary hydrated nitric acid yields, on ebullition, vapours which, when heated to dull redness, are decomposed into a similar mixture of red fumes and colourless oxygen. Fig. 63 shows an apparatus

Fig. 63.

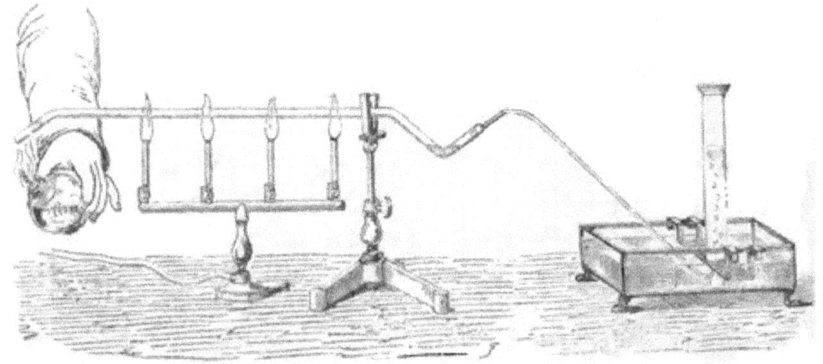

for performing this experiment. The nitric acid is boiled in the sealed and down-bent extremity of a tube, the continuation of which is kept red-hot by gas flames, so as to effect the decomposition of the vapours evolved. The same tube, further prolonged, conducts the gaseous products of decomposition to the trough, for collection of the oxygen over water; the elbow in this conducting portion of the tube serving to retain the acid liquors of condensation. In this experiment a compound, consisting of two volumes of nitrogen and five volumes of oxygen, abandons a certain proportion of its oxygen; and it is therefore obvious that, whatever may be in other respects the nature of the red fumes absorbable by water, they must at all events contain less than five volumes of oxygen in association with two of nitrogen.

The separation of this red fuming product from nitric acid

may be accomplished, even at ordinary atmospheric temperatures, by submitting the acid to the action of bodies possessing, like most of the metals, considerable attraction for oxygen. Tin, silver, copper, zinc, when immersed in nitric acid are powerfully attacked, with disengagement of gaseous products as before, but without escape of oxygen—this being appropriated by the metals which it converts into oxygen compounds termed oxides. The quantity of oxygen thus withdrawn from nitric acid varies with the nature of the metal used, and with the physical conditions of the experiment, especially with the temperature at which the operation is conducted, and the diluteness of the acid employed.

Tin, made to act on nitric acid under favourable circumstances, removes one-fifth of its oxygen with disengagement of a volatile compound, which at a low temperature condenses to white needles. This compound, which is termed *hyponitric acid*, contains two volumes of nitrogen united with four volumes of oxygen; or by weight, 28 of the former combined with 64 of the latter element.

When submitted under appropriate conditions to the action of silver, nitric acid loses two-fifths of its oxygen, and is converted into a yellowish-red vapour, which, on cooling, condenses into a bluish-green liquid. This substance is called *nitrous acid*, and contains two volumes of nitrogen combined with three of oxygen, or by weight 28 of the former to 48 of the latter.

Copper, when brought into contact with nitric acid, appropriates three-fifths of its oxygen with evolution of a transparent colourless gas, which consists of two volumes of nitrogen and an equal bulk of oxygen gas; or, by weight, 28 to 32 parts of the respective elements. This product, which is termed *nitric oxide*, has the remarkable property of acquiring a yellowish-brown colour by exposure to the air.

Under the influence of zinc, lastly, if the experiment be well conducted, nitric acid is deprived of not less than four-fifths of its oxygen, with evolution of a colourless gas, somewhat soluble in water, and which also remains colourless in contact with the air. This gas, which is called *nitrous oxide*, contains two volumes of nitrogen associated with one volume of oxygen, or by weight 28 parts of the former to 16 of the latter gas.

Thus, then, there are no less than five compounds of nitrogen with oxygen, the composition of which by volume and by weight, with the corresponding formulæ, is set forth in the following table :—

NAMES.	COMPOSITION.				FORMULÆ.
	By volume.		By weight.		
	Nitrogen.	Oxygen.	Nitrogen.	Oxygen.	
Nitrous oxide ..	2 vols.	+1 vol.	$14 \times 2 = 28+$	16	N_2O
Nitric oxide ..	2 vols.	+2 vols.	„	$28+16 \times 2 = 32$	N_2O_2
Nitrous acid ..	2 vols.	+3 vols.	„	$28+16 \times 3 = 48$	N_2O_3
Hyponitric acid	2 vols.	+4 vols.	„	$28+16 \times 4 = 64$	N_2O_4
Nitric acid ..	2 vols.	+5 vols.	„	$28+16 \times 5 = 80$	N_2O_5

The first thing that strikes us, on considering with attention this remarkable series of compounds, is the contrast they present, by their multiplicity, to the products obtained in all our previous experiments. As yet we have become acquainted with only one *chemical compound* of hydrogen with chlorine, of hydrogen with oxygen, and of hydrogen with nitrogen, though we could obtain *mechanical mixtures* of those gases in any proportions. We remained, therefore, within the limits of our then acquired knowledge, in considering chemical combination to be contradistinguished from mechanical mixture, by the single and immutable proportionality we had always seen the former affect, while the latter admitted manifold proportions.

In the presence of the phenomena we have just witnessed, this conception can evidently no longer be maintained.

If the five bodies just examined be chemical compounds, not mechanical mixtures, our former view of chemical combination must be modified and expanded to include them. And that they are true chemical compounds we have unequivocal proofs; first, in the definiteness and constancy of their composition; secondly, in the essential differences of property and character that distinguish them from their elementary constituents. Nitrogen and oxygen are colourless gases, insoluble in water, and incapable of condensation into the liquid (and *à fortiori* into the solid) form by any means at our command. Nitric anhydride and hyponitric acid, on the other hand, solidify at low temperatures to white crystalline bodies. Nitrous acid condenses by cold to a yellow-brown liquid; nitric oxide in contact with air acquires a bluish-green colour; nitrous oxide, lastly, is much more soluble in water than either of its constituents. It is therefore obvious that these bodies are not mere mechanical mixtures, but true chemical compounds of oxygen and nitrogen; and we are thus irresistibly led to the conclusion—one evidently of the highest importance—that two of the elements are capable of combining *in several proportions* to form a series of true chemical compounds, each differing from the others, and all differing from their primary constituents.

The difference between a mechanical mixture and a chemical compound does not, however, on this account become less sharply marked. In mechanical mixtures the elements may associate in proportions, whose name is legion, and which may be multiplied to any extent by arbitrary increments of this or that ingredient in the mass. In chemical compounds, on the contrary, the elements are united in comparatively few proportions; those which succeed to the first being all referrible thereto, as its multiples, in definite ratios. On the one hand limited, on the other boundless variety; and on the one hand well-defined, on the other indeterminate proportions. The possible *mixtures* of nitrogen with oxygen are beyond the power of numbers to express; the possible *combinations* of nitrogen with oxygen are only five; two volumes of nitrogen associating *chemically*, as we have seen, either with one, two, three, four, or five volumes of oxygen, and with no other proportions whatsoever. And should we ever succeed in producing other true compounds of nitrogen with oxygen, all chemical experience assures us that such combinations will take place in equally determinate ratios.

And herein, again, our study of the compounds of nitrogen with oxygen affords us a further lesson of the deepest interest. Not only is the number of these compounds limited, but the individual compounds, as far as their composition is concerned, bear to each other relations of the utmost simplicity. This simplicity is conspicuously displayed in the volume column of our table of the nitroxygen compounds. The volumetric proportion of oxygen united with two volumes of nitrogen is observed to advance in multiple proportions; the second, third, fourth, and fifth compounds containing respectively a double, treble, quadruple, and quintuple bulk of oxygen, as compared with the volume of this element in the first. There is nothing indeterminate in this series, as in a series of mechanical mixtures. The nitrogen volume being taken as constant in all the compounds, the oxygen volumes rise by definite bounds; and, as with volumes, so of course with weights. In the second column of the table we see the rise of the oxygen weight *per saltum*, in multiple ratios, throughout the series; and these definite proportions, both of volume and weight, are jointly expressed in the formulæ of the last column.

These facts, though gathered from the study of a single series of compounds only, illustrate one of the most general and comprehensive laws of chemistry; the master-principle which, as we shall hereafter find, connects and co-ordinates its simplest and most elementary truths with their furthest and most intricate developments.

The combinations of chlorine with oxygen form a series analogous to that of the compounds of nitrogen with oxygen just examined; while hydrogen and oxygen, which as yet we have seen associated only in water, form a second compound, called peroxide of hydrogen, in which a given amount of hydrogen is united with twice as much oxygen as is therewith associated in water.

These observations, however, by no means justify us in assuming that each couple of elements *necessarily* combines in more than one proportion. On the contrary, hydrogen, so far as our present information goes, combines with chlorine in one proportion only; that, namely, which produces hydrochloric acid. And so, again, hydrogen forms with nitrogen no other combination than that which results in the production of ammonia.

The composition of the five nitroxygen compounds, which has already so largely expanded our knowledge, still claims our brief attention from one other point of view.

The assumption of a constant amount (by weight or volume) of nitrogen in all these compounds is well calculated to exhibit the unbroken regularity of the steps by which the proportion of oxygen advances from the beginning to the end of the series. But this mode of representation is, on the other hand, disadvantageous in so far as it does not assign the simplest possible expression for two compounds of the series. On glancing at the column of formulæ in the table (repeated here for convenience)—

$$N_2O \qquad N_2O_2 \qquad N_2O_3 \qquad N_2O_4 \qquad N_2O_5,$$

we perceive that for the second and fourth formulæ, N_2O_2 (nitric oxide), and N_2O_4 (hyponitric acid), the simplest expressions would be obtained by halving each respectively: thus

$$\frac{N_2O_2}{2} = NO$$

is the simplest expression for nitric oxide; and

$$\frac{N_2O_4}{2} = NO_2$$

is the simplest expression for hyponitric acid. Thus corrected, the series of formulæ representing the nitroxygen series assumes this shape:—

$$N_2O \qquad NO \qquad N_2O_3 \qquad NO_2 \qquad N_2O_5.$$

Nor is their simplicity the only advantage resulting from this modification of the two formulæ in question. On inquiring into the *product-volume* thus expressed, in each of these two cases, we find that it coincides exactly with the *product-volume* expressed by the formula N_2O representing the first term of the nitroxygen series; and we find, further, that this product-volume coincides exactly with our normal product-volume, as determined in the typical cases of hydrochloric acid (HCl), of water (H_2O), of ammonia (H_3N), and of marsh-gas (H_4C). The product-volume we remember was, in all these cases, double the hydrogen unit-volume. This normal product-volume careful experiment has proved to be the actual product-volume resulting from the combination of nitrogen and oxygen to form N_2O, NO, and NO_2 respectively.

Having thus shown that three out of the five terms of the nitroxygen series obey the same law of condensation as that observed in our typical series of bi-elementary compounds, it remains to inquire whether the same normal condensation ratio obtains in the two remaining terms, N_2O_3 and N_2O_5. The strongest analogies justify us in assuming that these bodies conform to the law; but, owing to the special difficulties attending their manipulation, the respective product-volumes of these two compounds have not yet been determined by actual experiment.

It is therefore necessary, in the present state of science, to distinguish the first, second, and fourth terms of the nitroxygen series from the third and fifth, in respect to the evidence which we possess as to their condensation ratios. This we may best accomplish by drawing in *full lines* the squares employed to express volumes determined by experiment, and using (as we have done before) *dotted lines* to represent volumes at present hypothetical, though assumed on the ground of analogies which render their correctness in the highest degree probable.

This much premised, we may now graphically symbolize the nitroxygen series, both as to the volumes of its constituents and as to the product-volumes resulting from their condensation in combining, as follows:—

VOLUMETRIC SYMBOLIZATION OF THE NITROXYGEN SERIES.

Nitrous Oxide.

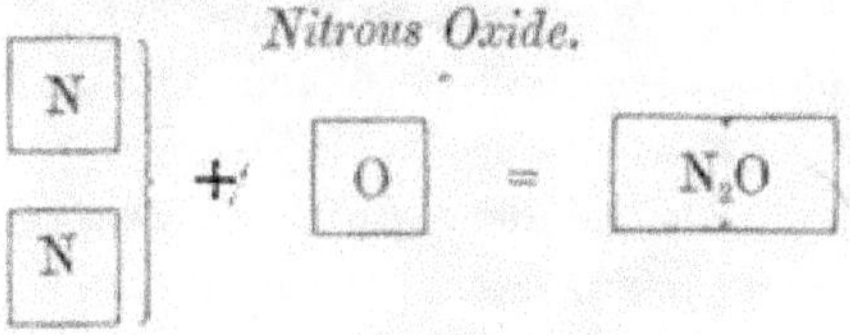

N
N
+ O = N_2O

Nitric Oxide.

N + O = NO

Nitrous Acid.

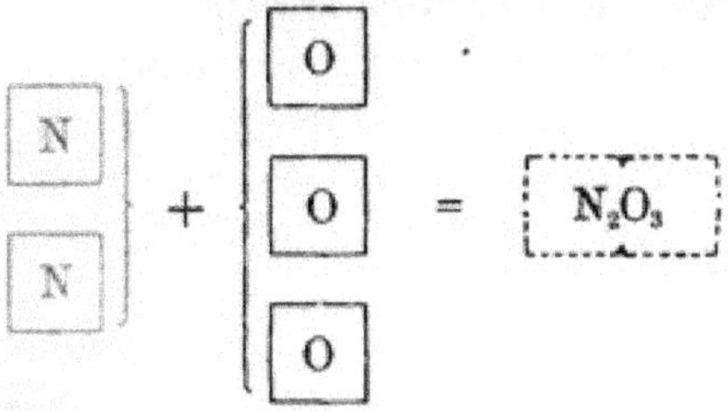

N
N
+
O
O
O
= N_2O_3

Hyponitric Acid.

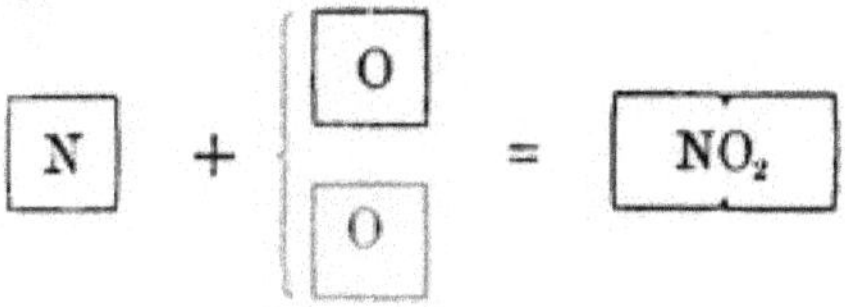

N +
O
O
= NO_2

Nitric Acid.

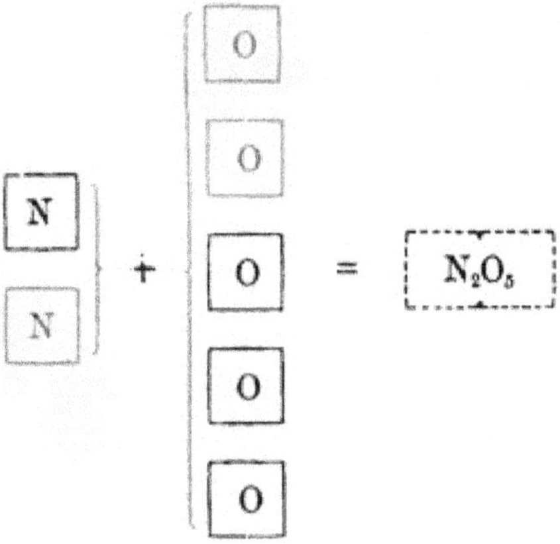

N
N
+
O
O
O
O
O
= N_2O_5

On comparing the condensation ratios shown in this table with those exhibited respectively in the symbolic diagrams of our typical compounds (comp. pp. 70 and 106), we see that the first and fourth terms of the series, N_2O and NO_2, though conversely constituted, are counterparts in condensation, both exemplifying the condensation ratio ($\frac{2}{3}$) which we found to obtain in the case of water (H_2O). The second term (NO) exemplifies combination without condensation ($\frac{1}{1}$), as typified by hydrochloric acid (HCl). The third and fifth, N_2O_3 and N_2O_5 respectively, exemplify the more complex condensation ratios, $\frac{2}{5}$ and $\frac{2}{7}$, for which our previous experience affords us no precedents, but whose complexity, I may mention in passing, we shall find very much exceeded in numerous cases which will hereafter come under our notice. These, however, are the two terms symbolized by dotted squares, in token of their still hypothetical character; and great, indeed, would be my pleasure if any of those assembled here should enable me to do away with these dotted lines, and make these symbols full ones like the rest.

With these reserves as to the third and fifth terms of the series, we may summarise as follows our present experience of the

Volumetric Composition and Condensation of Bi-elementary Chemical Compounds.

Constituents.	Product.	Condensation.
1 vol. + 1 vol. =	2 vols.	0 .
1 vol. + 2 vols. =	2 vols.	$\frac{2}{3}$
1 vol. + 3 vols. =	2 vols.	$\frac{1}{2}$
2 vols. + 3 vols. =	2 vols.	$\frac{2}{5}$
2 vols. + 5 vols. =	2 vols.	$\frac{2}{7}$

LECTURE VIII.

Transition from abstract to concrete formulæ—choice of a system of weights and measures to supply units for the expression of concrete values—obstacles thereto, as also to the diffusion of science generally, by the want of a universally-accepted system of weights and measures—French metrical system—reasons for adopting it—its general characters—principles of its nomenclature—comparison with English measures—the hydrogen litre-weight, or *crith*—the volume-weights of elements or compounds = the absolute weights of 1 litre at 0° C. and $0^{m}\cdot76$ pressure, expressed in *criths*.

Thus far, in our study of the weight and volume ratios traceable in the constitution of chemical compounds, we have referred all our figures to hydrogen as unity, without complicating our investigation by selecting some definitely-measured portion of space to serve as a standard volume, and assigning to each elementary and compound gas its actual density, as determined by weighing that volume in the balance.

To this inquiry, which will give to our formulæ, hitherto somewhat abstract, a more concrete, and therefore more practically useful, import, it is fitting that we should now proceed. And to this end it is in the first place incumbent on us to make choice of some one, among the many systems of weight and measurement in use, to supply us with the needful standards of gravity and dimension for reference.

And here I am fain to quit for a few moments my main topic, and open a parenthesis, which shall be as brief as possible, for the purpose of impressing upon your minds the obstacles presented to the cultivation and diffusion of scientific knowledge, by the want of one universally accepted system of weights and measures throughout the civilized world; and by the unfortunate prevalence, in lieu thereof, of a countless and confused variety of special weights and measures, differing not only for every nation and for every branch of human pursuits, but even for each nation in its different provinces, and often in its petty towns and

G

villages; so that not only are the standards of measurement multiplied, so as to fill bulky volumes with their mere enumeration, but that standards, nominally the same, such as, for example, the *pound*, the *foot*, the *bushel*, &c., have each of them in different localities different values, counting by hundreds, and probably, if all could be collected, by thousands.

It would be difficult to exaggerate the obstruction thus interposed in the way of those who desire, by extensive inductions, to collect and compare the experience of several nations, or the statistics of several localities, so as to enlarge the basis of facts on which all true science is founded.

Even to collate and compare the results of research in only two or three countries, such as England, France, and Germany, using different weights and measures, involves a perpetual labour of reduction, and a gratuitous absorption of time and force, deplorable when it is incurred, and still more deplorable when it operates, as it too often does, to make the records of one nation practically a sealed book to the philosophers of another.

I must not, however, dilate on these evils—the mere cursory indication of which is a digression from our immediate inquiry. Enough has been said, if the wish has been inspired to terminate as soon as may be the existing confusion, by promoting the universal adoption of a Unitary System of weights and measures.

This leads us to inquire Which, among the numberless systems in use, best deserves universal adoption? or, in other words, and to return to our more immediate subject, From which system should we select our standards of volume and density, for the purpose of assigning their concrete value to our hitherto abstract symbols and formulæ?

To this question, on which, it must be admitted, considerable differences of opinion prevail, my own answer is unhesitatingly given in favour of the French metrical system; the completeness and simplicity of which has already won it extensive recognition among the scientific communities of Europe, and incipient adoption by several enlightened nations as their legal standard for the ordinary purposes of life. It is, I believe, towards the

French metrical system that the balance of philosophic opinion strongly inclines throughout Europe; and this marked preponderance is in itself a cogent argument, where unity of volition and universality of practice are the very objects to be achieved.

I propose, therefore, in this place, to lay before you a concise sketch of the French metrical system, which, as it seems destined to come into general use throughout the world, so it has ever been, and will continue to be, our standard of reference throughout the whole series of chemical inquiries which we have undertaken together.

The French metrical system is developed in all its ramifications, whether of linear, superficial, or solid dimensions, whether of weight or capacity, from one primary element—that fundamental unit being a measure of *length;* and that standard measure being sought in the simplest and sublimest of human sciences, viz., in astronomy. This lineal unit is the 40-millionth part of the compass of our planet, as measured by a circumference passing around it in a plane including its axis. The circumference of the earth, thus taken, is termed a meridian, or "great circle," and its 40-millionth part has received the simple and well-chosen name *metre* (from the Greek *μέτρον, measure*). From this single term, as from a common root, all the ramifications of the French system—well, therefore, termed *metrical*—are derived. Prolonged by decimal multiplication, shortened by decimal subdivision; in other words, multiplied or divided by 10, by 100, by 1000, and so forth, the metre supplies all the degrees of *linear* measurement, from those required in the finest microscopic research to those employed to span the firmament and denote the mighty courses of the stars. Must we not, at the very outset, admit a certain grandeur in the conception of a system, whose encyclopædic scale supplies common and comparable expression for the dimensions of the minutest atom, and those of interstellar space?

As the *lineal* metre, decimally prolonged or shortened, furnishes the universal linear scale, so the metre *squared*, and that square decimally multiplied or divided, supplies all the gradations of *superficial* measurement; from the minute spaces of the philosopher's micrometer, to those larger ones marked on the land-surveyor's map, and to those still vaster areas in which the geographer estimates the expanse of continents.

So, again, the square metre, simply cubed, and that cube decimally multiplied and divided, supplies every gradation, from minutest to mightiest, of the cognate measures of *solidity* and *capacity*; in other words, of *space*, considered as *full* or *empty*. *Multiply* the cubic metre by one million, and you have a fit measure in terms of which to express the *capacity* of the Atlantic, or its *cubical contents* of brine; *divide* the cubic metre by one million, and you arrive at the petty volume of the gambler's ordinary *die*.

This last-named volume, the millionth of the cubic metre, taken as so much distilled water, furnishes the metrical *unit of weight*, called the *gramme*—a transition how admirable in its simplicity!—how useful, also, in forming, as it does, a link between the *volumetric* and *ponderal* appreciations of material quantity, in supplying for these different values cognate numerical expressions, in bringing them thus within the reach of easy, integral comparison (as contradistinguished from comparison by laborious fractional computations), and in facilitating, by these various means, the resolution of all the problems, theoretical and practical, which nature presents for solution by man!

The *gramme*, decimally multiplied and divided, forms, in this grand and simple system, the universal scale of weight. Its *millionth* does not turn the finest balance; its *millionfold* is the unit for heavy merchandise (like the English ton); with its *thousandth* commonly works the chemist; with its *thousandfold* the retail trader. The astronomer, when his business is to weigh this or that fixed star, has only decimally to multiply the millionfold of the little gramme in order to derive from it a weight-unit fitted for his use. Thus, by the adoption of this unitary scale, are the majestic oscillations of the celestial orbs rendered directly comparable with the vibrations of the chemist's granule in his balance.

From this general sketch of the French metrical system, we must pass to take cognizance of its details, and particularly to study the principles of its nomenclature.

These are as simple, and in their way as admirable, as the system itself. It is only necessary to bear in mind the names of the various units of Linear, Square, Cubic, and Ponderal quantity; as also to recollect that decimal prefixes taken from the Greek are employed to signify *multiplication*, while decimal prefixes taken from the Latin are used to express *subdivision*; and then, with these simple elements, the whole system, in all its details, can be built up and mastered in a few minutes.

The Greek prefixes for 10, 100, and 1000, are respectively *deca, hecto, kilo.*

The Latin prefixes for 10, 100, and 1000 are respectively *deci, centi, milli.*

To the word *metre*, the unit of Linear dimension, let us apply, first, the Greek prefixes which imply *multiplication*, secondly, the Latin prefixes which imply *subdivision.*

By the first process we get the following series :—

LINEAR MEASURE.

Unit—1 *Linear Metre.*

1. Metre-*multiples*—

		Metres.
Metre	=	1
Decametre	=	10
Hectometre	=	100
Kilometre	=	1000

By the second process we obtain the series :—

2. Metre-*divisions*—

		Metre.		
Metre	=	1		
Decimetre	=	0·1	or	$\frac{1}{10}$th of a metre.
Centimetre	=	0·01	or	$\frac{1}{100}$th ,,
Millimetre	=	0·001	or	$\frac{1}{1000}$th ,,

We have thus, in the Greek series, suitable names for tenfold, a hundredfold, and a thousandfold the unit; while the Latin series gives us appropriate appellations for its tenth, its hundreth, and its thousandth part.

Of the Greek series, the first and last terms (the metre and kilometre) are chiefly in use; the former for such purposes as the English yard subserves, the latter as road measure, in lieu of the English mile. The intermediate terms are but little required.

Of the Latin, or subdivisional series, on the contrary, all the terms are in constant use, replacing respectively the English yard, foot, inch, and various subdivisions of an inch (such as its 8th, 10th, 12th, 16th, &c.).

Each of the other sorts of measures, to wit, Surface measure, large and small, Cubic measure, Capacity measure, and Weight measure, is shaped out on precisely the same model; a convenient unit of quantity being, in each case, selected to start from.

Thus, for Surface measure, on the large scale (in use for measuring land), the metre square is deemed too small a unit to set out with, so its first decimal multiple, the decametre, is the linear unit selected for squaring, to obtain the primary surface unit; which is accordingly one square decametre, that is to say, a square of 10 metres in the side, or $10 \times 10 = 100$ square metres. This is called an *area*, or shortly an *are*. Applying to this unit, as before to the metre, first the Greek or multiplying, and then the Latin or subdividing prefixes, we obtain the following tables of metrical

SURFACE MEASURE (LARGE SCALE).

Unit—1 *Are*.

1. Are-*multiples*—

		Ares.		Square metres.
Are	=	1	=	100
Decare	=	10	=	1000
Hectare	=	100	=	10000
Kilare	=	1000	=	100000

2. Are-*divisions*—

		Are.		Square metres.
Are	=	1	=	100
Deciare	=	0·1	=	10
Centiare	=	0·01	=	1
Milliare	=	0·001	=	0·1

Of these measures the are and the hectare have come chiefly into use.

The centiare of this series, it will be observed, corresponds with the metre square; and this, with its decimal subdivisions, are employed for the measurement of *small* surfaces. Thus:—

SURFACE MEASURE (SMALL SCALE).

Unit—1 *Square Metre.*

Square metre-*divisions*—

		Metre square.
Metre square	=	1
Decimetre square	=	0·01
Centimetre square	=	0·0001
Millimetre square	=	0·000001

As the unit of weight, the choice of the French has (as already mentioned) judiciously fallen upon that die-sized cube already alluded to as resulting from the subdivision of the metre cube of distilled water (taken at 4°C., its point of greatest density) into a million equal parts. This they have called the *gramme*, from γράμμα, the name of a small weight in use among the Greeks. The word γράμμα is a derivative of γράφω, I write; and it probably came into use for this purpose, from the circumstance that the weights employed had their names, &c., inscribed upon them. Tabulating, as before, with the aid of the Greek and Latin prefixes, we obtain these two series:—

WEIGHT MEASURE.

Unit—1 *Gramme.*

1. Gramme-*multiples*—

		Grammes.
Gramme	=	1
Decagramme	=	10
Hectogramme	=	100
Kilogramme	=	1000

2. Gramme-*divisions*—

		Gramme.		
Gramme	=	1		
Decigramme	=	0·1	or	$\frac{1}{10}$th of a gramme
Centigramme	=	0·01	or	$\frac{1}{100}$th ,,
Milligramme	=	0·001	or	$\frac{1}{1000}$th ,,

All the terms of this latter, or divisional series are found useful; but only the kilogramme, in addition to the unit itself, is found requisite for ordinary use in the multiple series. The kilogramme, or thousand-gramme-weight, is the weight of the cube which results from dividing the metre cube into a thousand equal parts: in other words, from dividing the edge of the metre cube (*i. e.* the metre) into 10 parts (*i. e.* 10 decimetres) and cubing one of these parts (so as to obtain the decimetre cube). This same decimetre cube is also adopted as the unit for measures of capacity, in which function it loses the name kilogramme, and receives instead the appellation *litre* (from the term λίτρα, in use among the Greeks for the designation of one of their standards of quantity). From this unit of capacity, as the starting-point, the following capacity-measures are obtained by decimal multiplication and division, distinguished by Greek and Latin prefixes, as before:—

Capacity Measure.

1. Litre-*multiples*—

		Litres.
Litre	=	1
Decalitre	=	10
Hectolitre	=	100
Kilolitre	=	1000

2. Litre-*divisions*—

		Litre.		
Litre	=	1		
Decilitre	=	0·1	or	$\frac{1}{10}$th of a litre.
Centilitre	=	0·01	or	$\frac{1}{100}$th ,,
Millilitre	=	0·001	or	$\frac{1}{1000}$th ,,

All the measures in this table are employed alike for wet and dry bodies (wine, grain, &c.), and they are all more or less in use. The highest term (the kilolitre) corresponds to the metre cube, the unit of solid measure. The lowest degree of the table, or millilitre, is identical with the centimetre cube (*i. e.* the millionth of the metre cube, the die-like mass which, in distilled water, represents one gramme), and it is under this name that it is most commonly used.

It only remains to set forth the metrical cubic measures, which start, like the linear measures, from the metre cube, and form the two following series:—

Cubic Measure.

Unit—1 *Cubic Metre.*

1. Cubic metre-*multiples*—

		Cubic metres.
Cubic metre	=	1
Cubic decametre	=	1,000
Cubic hectometre	=	1,000,000
Cubic kilometre	=	1,000,000,000

2. Cubic metre-*divisions*—

		Cubic metre.
Cubic metre	=	1
Cubic decimetre	=	0·001
Cubic centimetre	=	0·000,001
Cubic millimetre	=	0·000,000,001

The cubic metre is in common use as a measure of wood used for fuel; in which capacity it takes the appellation *stere*, from the Greek *στερεός*, *solid*. From this, as from the other units of measure, a multiple and divisional series may be derived by aid of the Greek and Latin prefixes; but of the series so formed only one member, the *decistere*, or tenth of a stere, has proved needful, and come into use.

The highest term in each of the multiple series we have tabulated is that with the prefix *kilo*. This may, however, in each case, be decupled by substituting for *kilo* the prefix *myria* (also Greek). Thus a myriametre is equal to ten kilometres; a myrialitre to ten kilolitres, and so forth. These high expressions, however, are comparatively seldom required; and we have kept our tables simpler by reserving this, their final term, for separate mention, once for all, here at the close of our description.

It only remains to bring the units of this philosophical metric system into comparison with those of the English arbitrary scheme (the expression *system* would here be inappropriate) in order simultaneously to display, first, the marked superiority of the French over the English system; secondly, the arithmetical data for transforming the denominations of each into the other; and thirdly, the unwelcome amount of gratuitous toil attending such elaborate reductions. In the following table the French metrical units are placed in regard with, and reduced into terms of, their respective English correlatives:—

Metrical units.						
		Yard lineal.		Feet.		Inches.
1 metre	=	1·093633	=	3·280899	=	39·37079
		Perches.		Square yards.		Acre.
1 are	=	3·957388	=	119·603326	=	0·02471
		Grains.		Pound.		
1 gramme	=	15·434000	=	0·002204		
		Imp. pint.		Imp. gallon.		Cubic inches.
1 litre	=	1·760773	=	0·220096	=	61·027051
		Yard cube.		Cubic feet.		
1 stere	=	1·308020	=	35·316580		

For our present purposes attention should be particularly directed, first, to the units of weight and volume, namely, the gramme and the litre, in terms of which the weight and volume ratios with which we have to deal are best expressed; and, secondly, on the unit of length, the metre, in terms of which the barometric pressure is best expressed, the average height of the mercurial column being 0·76 metre = 76 centimetres = 29·9218 English inches.

Closing here our parenthetical account of the metrical system, and reverting to our immediate subject, we have now to select an appropriate *volume*, with its corresponding *weight*, to serve as our standard units of measurement.

Lineal Centimetres.

1	2	3	4	5	6

1 cubic decimetre,

= **1** litre,

= 1,000 cubic centimetres,

= $\frac{1}{1000}$th of a **cubic metre ;**

containing, at 4°C. temp^e^,

water, 1 kilo. = $\overset{\text{grammes.}}{1{,}000}$;

and, at 0°C., & $0^{m}{\cdot}76$ press^e^

hydrogen, 1 crith = $\overset{\text{gramme.}}{0{\cdot}0896}$

To face page 131.

After much consideration, I am disposed to select 1 cubic decimetre = 1 litre, as the most appropriate unit of volume for our purpose; and the weight of this measure of pure hydrogen (the body in all respects best fitted to serve as a standard) as our most suitable unit of weight. For this purpose hydrogen is taken at 0°C. temperature, and $0^{m}\cdot76$ pressure.

This volume-unit is represented by the cube shown on the opposite page; the front side of the cube being drawn of the true size, the other sides receding in perspective.

The actual weight of this cube of hydrogen, at the standard temperature and pressure mentioned, is 0·0896 gramme; a figure which I earnestly beg you to inscribe, as with a sharp graving tool, upon your memory. There is probably no figure, in chemical science, more important than this one to be borne in mind, and to be kept ever in readiness for use in calculation at a moment's notice. For this litre-weight of hydrogen = 0·0896 gramme (I purposely repeat it) is the standard multiple, or co-efficient, by means of which the weight of 1 litre of any other gas, simple or compound, is computed. Again, therefore, I say—Do not let slip this figure: 0·0896 gramme. So important, indeed, is this standard weight-unit, that some name—the simpler and briefer the better—is needed to denote it. For this purpose I venture to suggest the term *crith*, derived from the Greek word κριθὴ, signifying a barley-corn, and figuratively employed to imply a small weight. The weight of 1 litre of hydrogen being called 1 crith, the volume-weight of other gases, referred to hydrogen as a standard, may be expressed in terms of this unit.

For example, the relative volume-weight of chlorine being 35·5, that of oxygen 16, that of nitrogen 14, the actual weights of 1 litre of each of these elementary gases, at 0°C. and $0^{m}\cdot76$ pressure, may be called respectively 35·5 *criths*, 16 *criths*, and 14 *criths*.

So, again, with reference to the compound gases, the relative volume-weight of each is equal to half the weight of its product-volume. Hydrochloric acid (HCl), for example, consists of 1 vol. of hydrogen + 1 vol. of chlorine = 2 volumes;

or, by weight, $1 + 35{\cdot}5 = 36{\cdot}5$ units; whence it follows that the relative volume-weight of hydrochloric acid gas is $\frac{36{\cdot}5}{2} = 18{\cdot}25$ units; which last figure, therefore, expresses the number of *criths* which 1 litre of hydrochloric acid gas weighs at 0°C. temperature and $0^{m}{\cdot}76$ pressure; and the crith being (as I trust you already bear in mind) 0·0896 gramme, we have $18{\cdot}25 \times 0{\cdot}0896 = 1{\cdot}6352$, as the actual weight in grammes of hydrochloric acid gas.

So, once more, as the product-volume of water-gas (H_2O) (taken at the above temperature and pressure) contains 2 vols. of hydrogen + 1 vol. of oxygen, and therefore weighs $2 + 16 = 18$ units, the single volume of water-gas weighs $\frac{18}{2} = 9$ units; or, substituting as before the concrete for the abstract value, 1 litre of water-gas weighs 9 *criths;* that is to say, $9 \times 0{\cdot}0896$ gramme $= 0{\cdot}8064$ gramme.

In like manner the product-volume of sulphuretted hydrogen (H_2S) = 2 litres of hydrogen, weighing 2 criths + 1 litre of sulphur-gas, weighing 32 criths, together $2 + 32 = 34$ criths, which divided by 2 gives $\frac{34}{2} = 17$ criths $= 17 \times 0{\cdot}0896$ gramme $= 1{\cdot}5232$ gramme = the weight of 1 litre of sulphuretted hydrogen at standard temperature and pressure.

And so, lastly, of ammonia (H_3N): it contains in 2 litres 3 litres of hydrogen, weighing 3 criths, and 1 litre of nitrogen, weighing 14 criths; its total product-volume-weight is therefore $3 + 14 = 17$ criths, and its single volume, or litre weight is consequently $\frac{17}{2} = 8{\cdot}5$ criths $= 8{\cdot}5 \times 0{\cdot}0896$ gramme $= 0{\cdot}7616$ gramme.

Thus, by aid of the hydrogen-litre-weight or *crith* = 0·0896 gramme, employed as a common multiple, the actual or concrete weight of 1 litre of any gas, simple or compound, at standard temperature and pressure, may be deduced from the mere abstract figure expressing its volume-weight relatively to hydrogen.

From this knowledge, the weight of 1 litre of any gas, simple or compound, at any other than standard temperature or pressure, or under any variation both of standard temperature and pressure, may be deduced, by the application of the formulæ devised by physicists to express the laws of expansion and contraction, for gases under varying conditions of temperature and pressure. The volume is, for gases, inversely as the pressure; and every degree of temperature, added or subtracted, implies a certain fractional expansion or contraction of a gas's bulk. Diminished pressure and increased temperature consequently lighten the litre of any gas, in perfectly well-known and definite proportions. Hence, from the weight of 1 litre of gas, taken at any temperature, however high or low, the weight of 1 litre taken as at 0°C. can be unerringly deduced by computation.

This fact is of great value, as enabling us to include in common and comparable forms of expression the gas-volume-weights of bodies that are *not*, as well as of those that *are*, gaseous at 0° centigrade. Bodies which are liquid, or even solid, under this standard physical condition, have only to be weighed at the temperature, however elevated, at which they do assume the gaseous form, and from the weight of 1 litre, under this physical condition, the weight which 1 litre would assume at the standard temperature, could it remain gaseous thereunder, is deducible by computation.

The correctness of the weight and volume ratios thus obtained can be put to the test, conversely, by subjecting hydrogen to the same temperature as that which is needful to raise into gas the less volatile body with which we may desire to compare it. The physical conditions of attenuation being thus rendered equal for both bodies, their respective volume-weights may be experimentally determined, and the ratio of these recorded. In this way hydrogen, bromine, and iodine have been compared as to their relative volume-weights, at the temperatures needful to convert the two latter bodies into gas; and the values thus obtained have been found to stand to each other in the ratios 1 : 80 : 127 for hydrogen, bromine, and iodine respectively. In other words, to whatever volume the litre, or *crith*, of hydrogen may be dilated, that same volume of bromine and iodine, subject to like conditions of dilatation, will weigh 80 criths and 127 criths respectively.

Again, bodies which, like carbon, cannot be volatilized alone under any physical conditions whatsoever, but which become gaseous on combining with certain aeriform bodies—as, for instance, in the case of carbon, on combining with hydrogen—may be brought hypothetically under our general volumetric and ponderal expressions, by computations, based on analogy, and affording, not indeed certain, but more or less probable results. Thus 2 litres of marsh-gas are found to contain four litres of hydrogen, which, according to analogy, may be probably combined either with 1 litre of carbon gas, or, as in the exceptional cases of phosphorus and arsenic, with half that volume. As the weight of the carbon combined with 4 litres = 4 criths of hydrogen is known by analysis to be 12 criths, the litre of carbon gas, at standard temperature and pressure, probably weighs either 12 or 24 criths, accordingly as its deportment is assumed to resemble that of the majority of volatile and volatilisable bodies, or to coincide with the exceptional behaviour of phosphorus and arsenic. The former assumption has been adopted by many chemists for the purpose of provisionally assigning to carbon a place in the general system of volumetric and ponderal ratios; but this view, it must be borne in mind, is entirely speculative.

With these additions to our stock of knowledge, the symbols of the volatile elements, and the formulæ of the volatile compounds, acquire for us a new significance. We have already seen that they express the single volume-weights of all the elementary bodies symbolized, with the exception of phosphorus and arsenic, of which they express the half volume-weights only. We have also seen that, of the compound bodies, they express, in all cases, without exception, the product-volume-weights, so far as these have as yet come under our notice. We now find that, with the aid of one co-efficient, the key to them all, these abstract weights may be transformed into the corresponding concrete or actual weights at 0°C. temperature, and 0m·76 pressure. That key, or co-efficient, is the *crith* = 0·0896 gramme, and I think you will now see the grounds of my often-repeated hope that you will firmly commit it to memory.

TABLE OF GAS-VOLUME-WEIGHTS.

1	2	3	4
NAMES.	Symbols and Formulæ.	Gas-vol.-weights, or weights of one litre at 0°C. and 0m·76 Bar. = sp. gr. relatively to hydrogen in terms of a crith.	Gas-vol.-weights, or weights of one litre at 0°C. and 0m·76 Bar. = sp. gr. relatively to hydrogen in terms of a gramme.
SIMPLE BODIES.			
Hydrogen	H	1	0·0896
Chlorine	Cl	35·5	3·1808
Oxygen	O	16	1·4336
Nitrogen	N	14	1·2544
COMPOUND BODIES. *Typical.*			
Hydrochloric acid	H Cl	18·25	1·6352
Water-gas (dry steam) ..	H_2O	9	0·8064
Ammonia	H_3N	8·5	0·7616
Marsh-gas	H_4C	8	0·7168
Nitroxygen Series.			
Nitrous oxide	N_2O	22	1·9712
Nitric oxide	N O	15	1·3440
Nitrous acid	N_2O_3	38 ?	3·4048
Hyponitric acid	N O_2	23	2·0608
Nitric acid	N_2O_5	54 ?	4·8384

In the above table, the first column contains the names of

the simple bodies and their compounds, so far as our studies have yet extended; in the second column are placed their symbols and formulæ; in the third, their gas-volume weights, or specific gravities relatively to hydrogen at standard temperature and pressure, read in terms of one *crith;* while the fourth and last column gives the same values expressed in terms of one *gramme.*

For many purposes, where closely-approximative results suffice, trouble in multiplying may be saved, by employing 0·090 gramme instead of 0·0896 gramme as the value of the *crith.* The error for chlorine (the heaviest of the elementary gases) is only 3·1950 − 3·1808 = 0·0142 gramme in excess on the weight of the litre; for hydrogen (the lightest body), it is only 4-ten thousandths of a gramme.

I will only ask your attention to-day to one further remark on this subject, having reference to a point of difference worth noting between the symbols of the elementary and the formulæ of the compound gases. Read in *criths,* the former expressions give the weights of 1 litre, the latter of 2 litres, of the elementary and compound gases respectively; so that we may properly term the former *monolitral,* and the latter *dilitral,* expressions. In the exceptional cases of phosphorus and arsenic, the symbols, read in *criths,* give the weight of half volumes only of their respective vapours; so that these are *hemilitral* expressions. These discrepancies are attended with some inconvenience, which, however, I may mention in passing, the course of our future inquiries will enable us, in a great measure, to obviate.

The unit *crith* might, of course, be multiplied and divided by the Greek and Latin prefixes like any of the units of the metrical system; but no useful purpose would be served thereby. Indeed, the term *crith* may be dispensed with altogether by those whose memory can retain, without assistance, the value of the co-efficient 0·0896.

LECTURE IX.

Philosophical conceptions of chemical phenomena—hypotheses—theories—matter—its nature and essence—its conditions, solid, fluid, gaseous—its activities, molar and molecular—properties of molecules—their mutual cohesion and repulsion—nature and properties of gases—their elasticity—their latent heat—nature of heat—molecular dynaspheres—observed influence of temperature and pressure on gases—identical comportment of all gases under like variations thereof—composite structure of molecules—conception of atoms—how arrived at—final term of the known threefold divisibility of matter—its infinite divisibility, why not to be affirmed—elementary molecules—diatomic molecules—tetratomic molecules—their symbolization—atomic and molecular forms of notation—molecular symbolization exemplified in the formulæ of the nitroxygen series—comparative advantages of the atomic and molecular form of notation.

At the present stage of our inquiry, a new aspect of chemical phenomena presents itself to our notice, and its due consideration will justify us in assigning to the symbols of the elements, to the formulæ of the compounds, and to the equations expressing their reactions, a deeper signification than any they have hitherto possessed for us.

Up to the present moment we have not quitted the domain of experience; we have confined our attention to facts, either witnessed by ourselves, or accepted on the testimony of others. We have been satisfied to observe chemical phenomena, without seeking to explain them, save in so far as their orderly collection, comparison, and record may be held to constitute explanation.

To the causes of the remarkable effects we witnessed, our attention has not yet been turned. Yet the inquiry into the causes of observed phenomena is urged on us by one of the strongest instincts of our intellectual nature. That instinctive curiosity cannot, indeed, be fully satisfied. The first causes of phenomena lie beyond the limited scope of our perceptive and reasoning faculties. The conditions of their existence or production, and their relations of succession and similitude are, indeed, open to investigation; but their intimate nature and prime origin are for us inscrutable mysteries. We may, however, by the aid of imagination, form *hypotheses* (a word of Greek origin, from ὑπό, under, and θέσις, a derivative of *τίθημι*, I place, corresponding to the Latin *suppositio*, from *sub*, under, and *positio*, a derivative of *pono*, I place, whence the English word *supposition*) to connect the results of our experiments and to guide the course of our inquiries. And, though merely speculative hypotheses, dissevered from experimental investigation, are to be deprecated as vain and sterile exercises of ingenuity, hypotheses based upon facts, assisting in their conception, and deriving probability from the number thereof which they connect and explain, besides (and above all) tending to suggest new experiments, deserve to rank among the most valuable aids to scientific research.

Hypotheses are, of course, to be held provisionally, subject to modification and abandonment, in so far as they may from time to time prove inconsistent with the results of further experimental research. On the other hand, when hypotheses embrace and explain extensive ranges of phenomena, when experiment confirms the results they foreshadow, when successive discoveries raise them higher and higher in the scale of probability, they lose more and more their provisional character, and gradually assume the name and rank of *theories* (from the Greek *θεωρέω, I contemplate*), till at last they come to be embodied permanently among the recognized doctrines of philosophy and science.

The observed phenomena of combination in definite proportions by weight and volume are susceptible of explanation by a theory in the highest degree probable and suggestive, which the experiments we have made together, and the symbols in which we have recorded their results, have prepared us to understand.

To this theory we shall now devote some attention, without, however, venturing too far as yet upon speculative ground.

In order to arrive at the theoretical conception in question, we must ask ourselves, What *is* matter? Of what parts is it composed? How are these constructed and held together? How comes the very same matter, water for example, to present itself sometimes in the solid form, as ice; sometimes in the liquid form, as the same ice when melted; sometimes in the gaseous form, as the same melted ice changed to dry steam by further heating? And, lastly, what happens to matter, what changes does it undergo, when its various elementary forms combine, as we have seen them, to produce bodies having properties wholly different from those of their constituents?

These and other analogous questions have occupied philosophers during many generations; and the briefest history of the innumerable controversies thus engendered would fill many volumes. Even the single preliminary inquiry—Is matter infinitely divisible, or does it consist of smallest particles, incapable of further subdivision?—has given rise to contradictory opinions and arguments, which merely to summarize would largely exceed the limits of time and space at our disposal.

Experience of the value of hypothetical conceptions, when confined within just limits, and of their sterility when pushed too far, has imbued the leaders of modern philosophy with a spirit of extreme moderation and reserve in their attempts to penetrate these deep secrets of nature.

Setting aside, in a similar spirit, the more transcendental speculations of philosophers upon the nature of matter, let us here select for consideration those hypothetical conceptions of its structure which seem best adapted to connect and explain the results of modern research; and which, by enabling us to comprehend the phenomena we have already witnessed, may also assist us in shaping the course of our further experimental researches.

Let us, for this purpose, consider the familiar body, *water*, into the nature of which our experiments have already given us some insight; and let us consider it in its three conditions, as ice, as fluid water, and as water-gas or dry steam, all which are here before us. What is the first thing that strikes us in looking at them?

The first thing that strikes us is, that ice, water, and steam manifest two sorts of activity—one exerted by masses of sensible magnitude, acting through measurable distances of space; the other operating between particles, and through intervals of space, so minute as to be incommensurable.

The attraction of mass for mass of matter, which we see manifested in the courses of the celestial bodies, in the movement of falling bodies, and in the pressure of bodies at rest upon the ground, exemplifies the first kind of activity. This is equally observable in the ice, in the water, and in the water-gas; for these all possess *weight*; a sensible mass of either reciprocates attraction with the earth, through measurable distances of space.

The Latin for mass is *moles*; and its modern diminutive, *molecula*, is employed to designate "a little mass," that is to say, a material particle of incommensurable minuteness; hence the reciprocal actions of minute particles through insensible intervals of space are distinguished as *molecular*. We may fairly therefore contradistinguish, by the epithet *molar*, the reciprocal actions of measurable masses through measurable intervals of space.

The means of mechanical comminution at our disposal, our grinding-mills, mortars, and the like, do not carry us beyond the *molar* subdivision of matter. However finely we might grind up this ice, for example, if we took care to keep the temperature below freezing-point, we should still have masses consisting of several molecules. For, our finest ice-powder would still consist of very small fragments of solid ice; and if, of this ice-dust, we took the smallest grain, we could, by applying heat, turn it into water, thus proving it to have *parts*, capable of separation, so as to be rendered moveable amongst each other. There is no instance of liquefaction resulting from the mechanical comminution of a solid body. Hence we take it as certain that the most impalpable product of mechanical pulverization is still a cluster of molecules.

We are thus enabled to distinguish in matter two kinds of divisibility, *molar* and *molecular*; the former being accomplished by *mechanical* means, and only resulting, even when pushed to its utmost attainable limits, in the production of a molecule-cluster or mass of sensible dimensions, which may be termed a *mole*; while the latter is accomplished by *physical* means (that is to say, by aid of physical forces, such as heat), resulting in the disruption of the masses or *moles* into their incommensurably minute constituent *molecules*.

The study of the reciprocal action of material *masses*, or *moles*, constitutes the science of *mechanics*; a science of the deepest interest, abounding in simple and admirable laws, with which, however, we are not at present concerned.

Turning to the consideration of *molecular* activities, of those which are distinguished by the incommensurable minuteness of the particles of matter, and of the intervals of space, between and through which they take place; and looking once more at the samples of matter before us—at our ice, our water, and our water-gas or steam; we are again, as before, struck with a contrast between two diametrically opposite kinds of activity, one conspicuously manifested in the solid ice, and called *molecular cohesion*, the other especially manifested in the water-gas, and termed *molecular repulsion*. The former force gives to solid bodies their tenacity; to the latter, gaseous bodies owe their extreme tenuity, and the free mobility of their molecules amongst each other.

In fluid bodies, here represented by our water, we observe these two forms of molecular activity balanced at an intermediate point. The molecules of fluids cohere with considerable force; as you perceive, when I dip this rod into the water, and take out a bunch of them, sticking to each other, and also to the rod, in the form of this pendent water-drop; but this cohesion is exceedingly feeble as compared with that of the similar molecules agglomerated in the solid form here in our block of ice. Again, the molecules of fluids are moveable amongst each other; as you notice when I shake the water in this vessel, when I agitate it with this rod, and when I pour it into this other glass; but their mobility is far inferior to that of the molecules of gas. In vain should we dip our rod into the gas to take up a drop of it; we should obtain no coherent bunch of gas-molecules, like our pendulous water-drop. And it is precisely to their superior molecular cohesion that fluids owe their inferior molecular mobility as compared with gases. Hence the property of fluids termed their *viscidity*, a property which varies greatly in different fluids, so as to render unequal (for example) their rate of flow through tubes; but from which all gases are absolutely free, their molecules tending rather to recede from each other than to cohere. The only property which, in this respect, at all assimilates gaseous to fluid bodies, is the tendency of the former to adhere in thin films to the surfaces of solid bodies dipped into them or otherwise brought into contact with them. In this way a gas may be said to *wet* a solid body just as a fluid does; but here the resemblance ceases. For, in the case of the fluid, the inter-molecular adhesion, or viscidity, causes the first film to attract another film, and this a third, and so on, till a coating of sensible thickness is obtained; whereas, in the gas, the original film attracted by the solid does not in its turn attract a second, but remains of insensible thickness. How its presence is made known we shall hereafter learn from interesting experiments. At present these phenomena only concern us, as marking the different behaviour of matter, accordingly as its condition is fluid or gaseous.

This difference of comportment is not surprising when we reflect how much greater are the intervals which separate the molecules of a gas—of our water-gas, for example—than those which intervene between the molecules of the same body in the form of ice or of water.

Ice and water differ very little in bulk; at and near the freezing-point, indeed, water occupies rather *less* space, weight for weight, than ice. But water-gas or steam, at 100° C., occupies 1689 times more space than water at the same temperature.

Hence it follows that the molecules of water-gas are separated by intervals which, though doubtless incommensurably small, are nevertheless 1689 times larger than those which separate the same molecules when reduced to the fluid condition.

For the student of volumetric chemistry the structure of *gases* has a special interest; and it is on this branch of the inquiry into the nature of matter that we shall here bestow our principal attention.

What is the nature of the intervals between the molecules of a gas?—are they empty space, or are they filled? and, if so, how, or with what are they filled? That they are not empty spaces we have very good reason to believe, on account of the powerful resilient property manifested by gases when forcibly compressed. Here, for example, is a moist bladder full of a gas, of common air—you observe how powerfully it resists my forcible endeavours to compress it; and, when I strike it with this mallet, you see, by the bounding back of the implement, the powerful resilience, or elasticity, which gas possesses.

But what is the nature of this elasticity or resilience—to what power or force is it due?

Several phenomena point to *heat* as its cause. Heat is the agent by which ice is made to pass, through the fluid, into the gaseous form; and, with every increment of heat, the elastic power of the ice-derived gas augments. This is true, not of water-gas only, but of all gases; and of this we can easily obtain illustration by heating our bladder full of common air: you perceive how much tenser it becomes as the gas absorbs heat: when I strike it as before the mallet, you observe, is more powerfully resisted; and, upon heating it still more, its elasticity overcomes the tenacity of the envelope, and it bursts with the loud report you hear.

Again, if I set over a gas-burner a capsule containing a lump of ice of the temperature of 0° C., the ice gradually melts, and becomes converted into water; but if I try with a thermometer the temperature of the water at the moment when the last particle of ice is melted, I find it still at 0° C. That the capsule has been absorbing heat we are sure; first, because of its position over the flame; secondly, because the ice it contains has melted.

What has become of the heat thus absorbed? and how comes it that the water produced, when heat melts ice, is nevertheless an ice-cold fluid?

If any one be tempted to doubt the absorption of heat by the capsule, on account of the fact that the water it contains is ice-cold, nothing is easier than to convince him of his error. For this purpose it suffices simply to leave the capsule, with its ice-cold contents, in its place over the flame; and the thermometer will soon bear witness to a continuous and rapid absorption of heat. The water, ice-cold at starting, grows hotter and hotter every minute. The rate of increase of the heat shows the heat-absorbing power of the capsule; so that, by noting how many minutes the ice takes to melt, we know how much heat was absorbed during the process. Meanwhile, the thermometer plunged into the water continues to rise, till at last it reaches 100° C.; and then, as we all know, the water boils.

If, now, we still leave the capsule in its place, it continues, of course, to imbibe heat at the same rate as before; but the water now once again ceases to increase in temperature. It merely "boils away," as we familiarly say; that is, it becomes progressively converted into water-gas, which escapes by diffusion into the air, till at last all the water disappears.

If we now leave the empty capsule over the flame, we have soon another proof of the heat it has been imbibing all this time; for its bottom speedily becomes red-hot.

Now, precisely as much heat as, in a given time, entered the capsule to make it red-hot, precisely so much, in each equal interval of time, entered the ice to turn it into water, and entered the water to turn it into steam. Yet the water was not hotter than the ice just melted, nor the steam than the water just on the point of volatilization. Evidently, therefore, heat enough to make the bottom of the capsule red-hot several times over has entered into the water and into the water-gas. Yet it has not made these red-hot. What has become of the heat thus absorbed and hidden; or, to use the scientific expression derived from the Latin, thus rendered *latent?* We see, concurrently with its disappearance, the molecules of ice loosen into fluid water, and the molecules of water take wing and form resilient gas. I could even give you, did time permit, experimental proof that, by condensing the water-gas back into water, we can recover the hidden force as sensible heat again; and, in like manner, that to freeze the condensed water so obtained into ice, there must be withdrawn from it just so much heat as the ice, by whose liquefaction we obtained it, absorbed and rendered latent in melting. Evidently then, putting these things together, we cannot help connecting the absorption and latency of the heat with the successive development of the fluid and gaseous conditions; nor can we regard the elasticity of the steam otherwise than as a result due to some form or modification of the force known to us as *heat.*

To the questions, therefore, what is a gas? and with what are the intervals between its molecules filled? succeeds the question, what is heat? This brings us face to face with one of the most ardently-mooted and deeply-interesting philosophical questions of the day. For some, heat is a species of thin ether, vibrating in the manner of light; for others, it is a pure force, having neither parts nor weight; for a third class of thinkers, of late years the majority, heat has no separate existence, but is merely a mode of motion, the result of the vibration of material molecules.

It is no part of our present task to attempt the solution of this deep and difficult problem. We may content ourselves here with the conception that heat, whatever may be its intimate nature, so operates, when it becomes latent in a gas, as to surround each molecule with a sort of repellent atmosphere which tends to keep it apart from its fellows; and that these molecular force-spheres—or, to employ the Greek equivalent, *dynami-spheres*, more shortly, *dyna-spheres* (from the Greek δύναμις, *force*, a derivative of δύναμαι, *I can*)—when mechanically compressed, counteract the pressure with exactly equal energy, and on the removal of the pressure, restore the gas (other things being equal) to the exact volume it previously possessed.

Now, carrying on observation side by side with hypothesis, let us compare with the behaviour of water-gas, under varied conditions of temperature and pressure, that of the other compound gases, hydrochloric acid gas, ammonia, and marsh-gas, and also that of our four simple or elementary gases, hydrogen, chlorine, oxygen, and nitrogen; or, to obviate the difficulty which attends the study of water-gas, on account of its falling back into the fluid state at ordinary temperatures, let us select, from among the more permanent compound gases, some type to represent them, say hydrochloric acid; and in like manner let us select from among the simple gases a typical representative, say hydrogen; and let us submit these two representative gases to equal increments and decrements of temperature and pressure successively. If they behave differently, we shall have reason to believe that the molecular structure of elementary and compound gases is different; if they behave alike, we shall be justified in assuming their molecular structure to be the same.

We have here an apparatus fitted up for the purpose of making this comparison rapidly, and on a scale easily visible. It consists of a modification of the double U tube, used already for a different purpose (comp. p. 53); and it has one long and simple limb, and one short limb bifurcated into two branches, each of which is provided with a stopcock. These short limbs are, moreover, enclosed in a cylinder of glass, as shown in the figure. The purpose of this glass envelope will appear in the sequel. Near the bottom of the apparatus another stopcock will be observed, so placed as to serve for emptying it. The whole apparatus being filled with mercury, the gases to be compared may be easily drawn, each into one of the two small branch-limbs, by connecting these, through flexible tubes attached to their stopcocks, with the apparatus supplying the respective gases, and then letting out the mercury through the stopcock below. Into the vacuum which the descent of the mercury

FIG. 64.

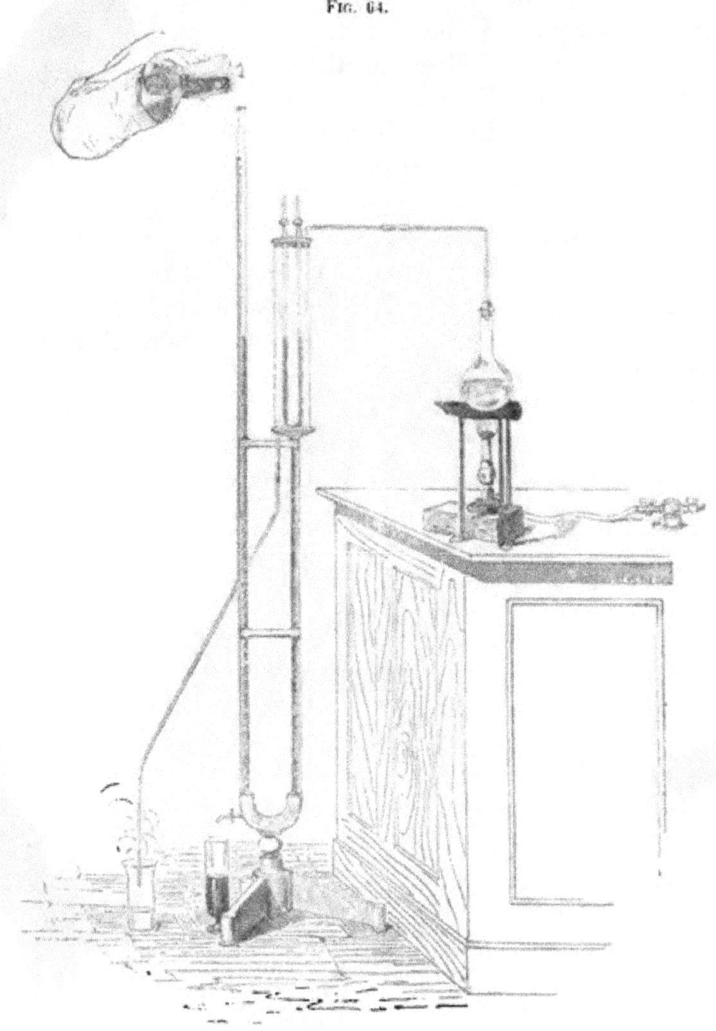

leaves in the branch-tubes, the gases to be studied of course flow; and, when the branches are thus about half filled, the stopcocks are closed; care however being taken to have a precisely equal quantity of gas in each of the two limbs. This adjustment made, we may mark with caoutchouc rings the space filled with gas in each tube. Upon pouring more mercury into the long tube, we obtain a column which exercises precisely equal pressure on the gases in the two branches; and we see that these are equally compressed, by the equal ascent of the mercury in each branch. On the other hand, on letting out mercury through the bottom stopcock, so as to lower the column and decrease the pressure, we observe that both gases, different as they are by nature, undergo equal dilatation. And so, again, the expansion and contraction of the two gases is shown to be equal, under equal increments and decrements of heat, by filling the cylinder which surrounds the bifurcated limbs of the apparatus with an atmosphere of hot steam or cold air by turns.

It thus stands clearly demonstrated that, if equal volumes of the elementary gas, hydrogen, and of the compound gas, hydrochloric acid, be taken under any given temperature and pressure, and the pressure be doubled for each, each becomes reduced to half its former volume, and at the same time acquires double its former resilient force, or elasticity; which it exerts in counterbalancing the pressure from without.

It stands equally proved that, if equal volumes of hydrogen, and of hydrochloric acid gas, taken at equal degrees of pressure and temperature, be exposed to equal increments or decrements of heat, they undergo equal degrees of expansion and contraction.

Experiments of this kind, repeated for numerous gases, simple and compound, have established as a law, that all true gases, simple as well as compound, comport themselves in sensibly the same manner under like variations of temperature and pressure; whence the inference fairly follows that their molecular structure is the same. Assuming, then, each gaseous molecule to be clothed or enveloped by a resilient dynasphere (as we have termed it), due, in some unknown way, to the influence of latent heat, experiment justifies us in inferring, from the identical comportment of all gases, when exposed to like variations of temperature and pressure, that they all contain, in equal volumes, an equal number of molecules so clothed; and that, as an obvious corollary, the diameter of these gas-molecules (including in that term as well the dynaspheres as their material nuclei) is, under like physical conditions, precisely the same for all gases. To express it more shortly, our unit-volume, or litre, whether of hydrogen, of hydrochloric acid, or of any other gas, simple or compound, is composed of mutually repellent dynaspheric molecules, equal (*omnibus paribus*) as to their *number*, and (consequently) as to their *size*.

At this point of our inquiry we may advantageously resume the consideration of material divisibility, of which we have already studied two forms or grades, the *molar* and the *molecular*; the former consisting in the mechanical disruption of large masses into small ones, the smallest still possessing sensible magnitude; while the latter is the further disruption, by physical agents, such as heat, of moles or masses, whether large or small, into their constituent molecules; that is to say, into parts contradistinguished from the minutest moles by the fact that they (the said parts) possess no commensurable magnitude at all. In the particular sample of matter which we selected for study, as being the most familiar of all compounds, we saw molecular succeeding to mere molar division, when heat melted comminuted ice into water, and then raised water into invisible steam or gas, by clothing its molecules with the mutually repellent dynaspheres, each dynasphere 1689 times larger than its material nucleus.

Infinitesimal as this subdivision of matter appears, inexpressibly minute as we cannot but conceive the material particles to be that form the central nuclei of the dynaspheres of bodies so attenuated and rare as the invisible gases, we yet know—we have experimental proof—that a further comminution of matter is possible; and that, as the smallest mass or mole of any compound may be broken up into its constituent molecules, immeasurably smaller still, so the ultimate molecule itself, however small we may choose to conceive it, is nevertheless still a *compound*, consisting of at least two parts, which, by chemical agency, may be detached from each other, so as to resolve the compound into its elements.

Here the divisibility of matter, so far as our knowledge, and the means of operating at our disposal, extend, reaches its final term. The *elementary* bodies are, as we remember, so called precisely because they resist every agency, mechanical, physical, and chemical, which we can bring to bear in the hope of dividing or decomposing them. We may imagine the two elementary particles which form the compound molecule of hydrochloric acid, for example, to be as small as we please. In this respect we may give the imagination free rein; we may conceive the particle of hydrogen, or of chlorine, to be divided and subdivided as many millions of times as we like, or rather, until the imaginative power is baffled by sheer exhaustion in the endeavour to push this conception further. No experiment yet made tends to restrict the freest range of our mental faculties in this direction; their only limitation lies in their own finite scope, doubtless more or less extensive in different minds. But, when we have, each of us, thus reached the idea of the smallest elementary particle which it is within the power of the mind to picture, all experience stands opposed to our going still further, and presuming to declare the elementary particles capable of division *ad infinitum.* Not one experimental result can be adduced in support of such an assertion. At this point, therefore, the experimental philosopher arrests his inquiry. Beyond this limit he sees only the dream-land of metaphysical speculation—a region essentially sterile because shut out from cultivation by means of experiment, from which alone can spring the harvest of Truth in the proper sense of the word; having for its foundation natural facts; for its object the study of their relations; for its result the determination of their laws.

To the metaphysical speculators, therefore, let us cheerfully resign the utterly futile and fruitless discussion whether even elementary matter may not be infinitely divisible. It is enough for us to know that, at all events, *we* cannot infinitely divide it; but that, relatively to our powers and purposes, to the limits of our imagination as well as of our experience, the assertion of the infinite divisibility of the elements is one we are not justified in making.

We thus arrive at the conception of indivisible particles as the ultimate constituents of elementary bodies, and these particles have received the appropriate name of *atoms* (from the Greek word τέμνω, *I cut, I divide*, with the privative α prefixed in token of negation).

The addition of this final term completes and enables us to epitomise our view of the threefold divisibility of matter, molar, molecular, and atomic; the first (molar) being performed by *mechanical* means, and resulting, when pushed to its utmost limits, in masses or moles (clusters of molecules) characterized by their possession of sensible magnitude; the second (molecular) accomplished by the agency of the physical forces (heat, electricity, &c.), employed under special conditions for the purpose, and resulting in the production of the dynaspheric *molecules* of which we reasonably conceive compound bodies to consist; the third (atomic) being capable of accomplishment only by agencies, such, and so applied, as to produce *chemical* decomposition, breaking up the incommensurable molecule itself into its elementary particles, which (as just explained) are called *atoms*, because incapable of further disruption or comminution by any means at our disposal.

This conception of the threefold divisibility of matter, molar, molecular, and atomic, being once clearly understood, and firmly grasped by the mind, we may usefully proceed, in the light which this theory supplies, to compare as to their structure compound with elementary gases. At first view we should be disposed, perhaps, to anticipate as probable, that, while the compound gases would be formed of divisible molecules or atom-clusters, the elementary gases would present no such complexity of structure, but consist merely of separate and indivisible elementary particles. But a little consideration will show us that this view is incompatible with the results of our preceding inquiry.

Confining our attention, as usual, to the gases with which we have made experimental acquaintance, let us select from among these, for the comparison in question, the simplest of our compound gases, hydrochloric acid, and its two elementary constituents, hydrogen and chlorine. And, that our views may be the clearer on this subject, which we shall find to be important, let us revert for a moment to our earliest experiments, and refresh our memories by reviewing once again the volumetric composition of hydrochloric acid gas.

Of this gas, we remember, the normal product-volume (2 litres) is formed by the association of one unit-volume (1 litre) of hydrogen with the same volume of chlorine gas; their union taking place without condensation. Each molecule of hydrochloric acid, therefore, is evidently constructed of two atoms at least; one being a hydrogen and the other a chlorine atom.

We have already gone over the experimental grounds for believing that equal volumes of all gases, whether simple or compound, contain, under like conditions, like numbers of dynaspheric molecules; which are, therefore, necessarily of equal size in all cases.

Let us now, to simplify our calculations, assign to the unknown number *n* of hydrochloric molecules, existing in our bilitral volume of hydrochloric acid gas, some definite numerical value, say 1000.

This being assumed as the number of molecules in 2 litres, the number in 1 litre is of course just half, or 500; and, as we recognise that equal volumes of all gases contain equal numbers of molecules, the litre of hydrogen and the litre of chlorine, which go to the formation of our 2 litres of hydrochloric acid gas, must likewise contain 500 molecules each.

Now, as each molecule of hydrochloric acid contains 1 atom of hydrogen joined to 1 atom of chlorine, the 1000 molecules of hydrochloric acid must, of necessity, contain 1000 atoms of hydrogen joined to 1000 atoms of chlorine—the whole number of atoms present being therefore 2000.

But we have just seen that one litre of hydrogen and one litre of chlorine contain, not 1000 molecules each of the respective bodies, but only 500.

It follows clearly that 500 *molecules* of hydrogen and 500 *molecules* of chlorine have supplied respectively twice as many *atoms* of those constituent bodies; each contributing its 1000 atoms to the aggregate number of 2000 atoms, existing in the 1000 hydrochloric molecules, contained in our 2 litres of hydrochloric acid gas.

If 500 molecules of an elementary gas supply 1000 atoms, it is plain that each molecule supplies 2 atoms; and thus we clearly perceive that the molecule of the compound gas under review, and the molecules of each of its elementary constituents, are all formed on the same type—that type being the first of our quadruple series, viz., the hydrochloric acid or *diatomic* type.

This is a remarkable and striking, yet strictly logical deduction. It completes a chain of reasonings which, if correct, justify the conception that simple as well as compound gases are complex as to their molecular structure; and that this structure, for hydrogen and chlorine, is of the diatomic type, also exemplified in hydrochloric acid.

Similar considerations show that this diatomic molecular structure characterises all the other permanent elementary gases with which we have made acquaintance.

These newly-attained truths give rise to the necessity of a new adaptation of our symbolic language for their expression. Hitherto the symbols of the elements have merely expressed for us their unit-volumes and combining weights; and the formulæ of the compound gases have, in like manner, only represented to us their product-volumes and combining weights. These expressions we have now to clothe with new meanings, corresponding to our new conceptions concerning atoms and molecules respectively.

To this new service we shall find our old formulæ lend themselves with admirable flexibility. It is true, no doubt, that our old expression for the unit-volume and combining weight of hydrogen (H = 1) conveys no adequate picture of the *diatomic* structure of its *free* molecule. But this symbol answers perfectly well to express the single atom of hydrogen as it exists in *combination*—

as, for example, we find it in the diatomic molecule (HCl) of hydrochloric acid. Again, the expressions employed to represent the product-volumes and weights of the compound gases require no modification to fit them for representing the atomic structure of the molecules of those gases. For example, the formula,

$$H + Cl = HCl = 1 + 35{\cdot}5 = 36{\cdot}5,$$

affords as perfect a picture of the diatomic molecule of hydrochloric acid as it does of its bivolumetric composition; at the same time accurately expressing the weight of its product-volume. And lastly, as we know that the molecules or atom-clusters of the free elementary gases are formed on the binary type of hydrochloric acid, we may evidently seek in the formula of that acid a model on which to frame appropriate symbolic expressions for those free elementary molecules.

Thus, for example, the molecular formulæ for free hydrogen and free chlorine are respectively—

$$\text{for hydrogen } H + H = HH = 1 + 1 = 2,$$

$$\text{and for chlorine } Cl + Cl = ClCl = 35{\cdot}5 + 35{\cdot}5 = 71.$$

Hence, it further appears that, for the volumetric symbolization of the free elementary gases, our formerly-used monolitral (and, in exceptional cases, hemilitral) expressions are no longer appropriate, but must be replaced by dilitral symbols like those employed to denote the normal product-volumes of the compound gases.

We thus obtain, as the correct molecular expressions for hydrochloric acid and its elements, the following bi-volumetric symbols:—

HCl = [HCl]

HH = [HH]

ClCl = [ClCl]

Upon these models the formulæ of the other elementary gases may be readily framed; and the equations representing the reactions in which they participate must of course undergo corresponding modifications. For example, the expression H + Cl = HCl, or, diagrammatically represented,

H	+	Cl	=	HCl

heretofore employed to represent the synthesis of hydrochloric acid by the direct combination of the two free gases, its constituents, is no longer, for us, a true representation of the facts; seeing that we know those two gases to be formed not, as from this equation would appear, by the reaction of solitary atoms, but by that of *diatomic molecules*. The true molecular equation of this synthesis becomes, therefore,

$$HH + ClCl = HCl + HCl,$$

or, diagrammatically represented,

HH	+	ClCl	=	HCl	+	HCl

or, again, reduced to its most succinct expression,

$$2H + 2Cl = 2HCl.$$

The dilitral symbolization of the elementary gases has a further advantage; it brings them into direct comparison, volume for volume, with the compound gases (also dilitral, as we remember), so that a list of the molecular weights of the elementary and compound gases represents also their relative volume-weights or specific gravities. We may express these either in terms of the unit-vol. of hydrogen,

H	=	1;

or, preferably, in terms of the product-volume of hydrogen,

H H	=	2;

and we thus obtain the two symbolic and numerical series which are placed side by side in the following table:—

Names of gases, elementary and compound.	Specific Gravities.	
	Monolitral. $\boxed{H} = 1$	Dilitral. $\boxed{HH} = 2$
Hydrogen	$H = 1$	$HH = 1 \times 2 = 2$
Chlorine	$Cl = 35{\cdot}5$	$ClCl = 35{\cdot}5 \times 2 = 71$
Oxygen	$O = 16$	$OO = 16 \times 2 = 32$
Nitrogen	$N = 14$	$NN = 14 \times 2 = 28$
Hydrochloric acid	$\frac{HCl}{2} = \frac{36{\cdot}5}{2} = 18{\cdot}25$	$HCl = 36{\cdot}5$
Water-gas	$\frac{H_2O}{2} = \frac{18}{2} = 9$	$H_2O = .. 18$
Ammonia	$\frac{H_3N}{2} = \frac{17}{2} = 8{\cdot}5$	$H_3N = .. 17$
Marsh-gas ..	$\frac{H_4C}{2} = \frac{16}{2} = 8$	$H_4C = ... 16$

It will be observed that, in the monolitral column of this table, the formulæ of the compound gases are halved, while in the dilitral column the symbols of the elementary gases are doubled, in order to render the whole series comparable in terms of the standard adopted in each case. Our preference is given to the second arrangement, because it distinctly shadows forth the true molecular structure of the free elementary gases, depicting clearly their analogy in this respect with the molecular type of hydrochloric acid. And, while this important chemical conception is figured in the four first expressions of the dilitral column, its four last members remind us of another and an equally significant chemical truth, viz., that the elementary gases, though in the free state they always exist as molecules, very commonly split up into separate atoms when they enter into combination. Illustrations of this fact, with reference to hydrogen, are supplied by the expressions HCl and H_3N in the table. For, though we might conceive the other two expressions, H_2O and H_4C to represent, the former the union of one hydrogen-molecule (HH), and the latter of two hydrogen-molecules (2HH) with oxygen and carbon respectively, we cannot but see in the formula HCl the representation of a single hydrogen atom (H = 1) associated with a single chlorine atom (Cl = 35·5); while, in the formula H_3N, we are equally constrained to recognize the representation of at least one separate hydrogen atom, namely the third,—that one which, in the expression HHH, is in excess of the diatomic expression HH, corresponding to the molecule of free hydrogen.

This consideration is of value, because it enables us to assign to the molecules and atoms of the elementary bodies contradistinctive definitions, quite irrespectively of any question as to the greater or less divisibility of matter. That is to say, we may define the *atom* of any given elementary body to be the smallest proportional weight thereof which is capable of existing in *chemical combination;* and we may define the *molecule* of an elementary body to be the smallest proportional weight thereof which is capable of existing in the *free or uncombined state.*

This definition of the elementary molecule, it will be observed, does not assume for it any particular numerical relation to the elementary atom; its terms are wide enough to admit the conception not only of diatomic, but also of triatomic, tetratomic, and polyatomic molecules; while, on the other hand, its terms do not exclude the conception of coincidence in weight between the atom, or *combining minim,* and the molecule, or *free minim,* of an elementary body; in other words, this definition admits as possible the conception of a *monatomic molecule.*

With the above models of molecular formulæ before us we shall have little difficulty in extending this mode of symbolization to the other elements which we have passed in review, viz., to the analogues of chlorine, bromine and iodine; to the analogues of oxygen, sulphur and selenium; to the analogues of nitrogen, phosphorus and arsenic; and lastly, to the analogues of carbon, silicon and titanium. In the cases of phosphorus and arsenic, we shall, of course, again encounter the same exceptional volumetric relations that obliged us to modify our former symbolization of these bodies; and we shall have to meet the anomaly by a corresponding modification of the molecular expressions. The following table presents a collective view of these elements, and of their molecular construction and symbolization :—

MOLECULAR CONSTRUCTION AND SYMBOLIZATION OF ELEMENTARY BODIES :—

Chlorine Group.

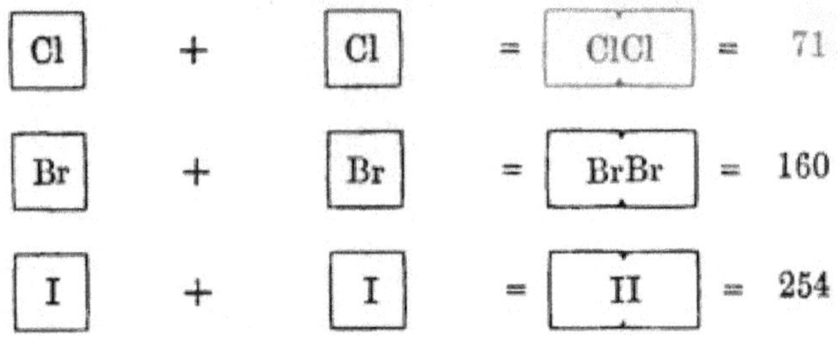

Cl	+	Cl	=	ClCl	=	71
Br	+	Br	=	BrBr	=	160
I	+	I	=	II	=	254

Oxygen Group.

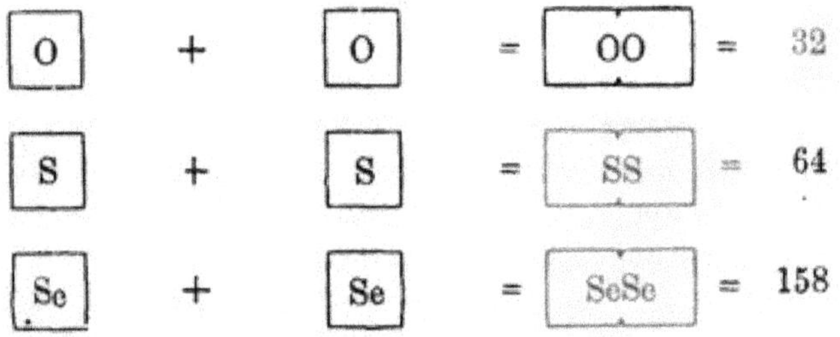

O	+	O	=	OO	=	32
S	+	S	=	SS	=	64
Se	+	Se	=	SeSe	=	158

Nitrogen Group.

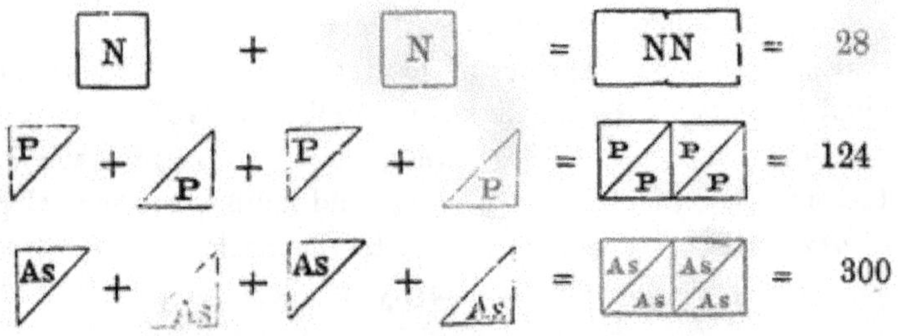

N	+	N					=	NN	=	28
P	+	P	+	P	+	P	=	P P / P P	=	124
As	+	As	+	As	+	As	=	As As / As As	=	300

Carbon Group.

C	+	C	=	CC	=	24 (?)
Si	+	Si	=	SiSi	=	57 (?)
Ti	+	Ti	=	TiTi	=	100 (?)

The symbolization of PPPP and AsAsAsAs in this table explains itself at a glance. The dilitral weight of these two bodies, that is to say, the weight of their molecule, or free minim, in the state of vapour, is four times as great as that of their atom, or combining minim, as determined in phosphoretted hydrogen, H_3P, and arsenetted hydrogen, H_3As. This, indeed, follows as a corollary from our previous observations on these bodies; for, as their unit volume, or monolitral volume, had to be represented respectively by PP, or 2P, and AsAs, or 2As, it is obvious that their dilitral or molecular formula must be twice as great; or $2P \times 2 = 4P$, and $2As \times 2 = 4As$. In these cases, accordingly, the molecule is tetratomic. In the carbon group we adhere, it will be observed, to the use of dotted lines in the diagrammatic symbols to imply that the expressions are hypothetical; and we add a note of interrogation within brackets to the written symbols for the same purpose.

With the molecular equation of the synthesis of hydrochloric acid as a model, we may readily reconstruct, in the molecular form, our old atomic equations representing the syntheses of other compound gases. It will suffice, therefore, to give one or two examples of such syntheses, expressed in molecular formulæ, and for this purpose we will select water and ammonia. Our old formula of the formation of water from its elements was

$$H_2 + O = H_2O;$$

for which we now substitute

$$2HH + OO = 2H_2O.$$

Similarly, should means be found to form ammonia directly from its elements, the atomic equation representing its synthesis would be

$$3H + N = H_3N;$$

for which, substituting the molecular expression, we have

$$3HH + NN = 2H_3N.$$

Once more: just as synthetic equations, adapted to the molecular hypothesis, exhibit, as starting-points of the reactions, molecules, not atoms, of the elements taking part therein; just so, conversely, molecular equations, when employed to represent reactions of decomposition, must be so constructed as to express in molecules, not in atoms, the elements set free.

Thus, the ordinary equations which, in a previous lecture, represented for us the decomposition of hydrochloric acid, water, and ammonia, by sodium (Na), were as follows:—

$$HCl + Na = NaCl + H.$$

$$H_2O + 2Na = Na_2O + 2H.$$

$$H_3N + 3Na = Na_3N + 3H.$$

The molecular equations representing these decompositions are—

$$2HCl + NaNa = 2NaCl + HH.$$

$$H_2O + NaNa = Na_2O + HH.$$

$$2H_3N + 3NaNa = 2Na_3N + 3HH.$$

In the former of these two series of formulæ it will be noticed that the first and third equations represent Na as used, and H as escaping, wholly or partly, in single atoms. In the corresponding equations of the second group, the expressions employed for free sodium and hydrogen represent them as consisting of atoms in couplets, *i.e.*, as diatomic molecules; a structure to whose type we know that hydrogen conforms, while, strong analogies, justify us in assuming a similar construction for free sodium gas. On comparing with each other the middle equation of each set, it will be seen that they are identical expressions; a circumstance due to the fact that, in the decomposition of water by sodium, the proportion both of the metal employed and of the hydrogen set free is 1 molecule = 2 atoms (comp. p. 218).

The two following series of formulæ set in like contrast the old and new modes of symbolizing the decomposition of water and ammonia by chlorine:—

Atomic Notation.

$$H_2O + 2Cl = 2HCl + O.$$
$$H_3N + 3Cl = 3HCl + N.$$

Molecular Notation.

$$2H_2O + 2ClCl = 4HCl + OO.$$
$$2H_3N + 3ClCl = 6HCl + NN.$$

In constructing molecular equations of compound bodies care must be taken to select those expressions which, while representing the true relative proportions of the elementary constituents, embody for this purpose the smallest number of atoms with which the compound or product-molecule can be built up. On the other hand, symbolic expressions must never comprise fractional parts of atoms: such fractional formulæ are of course inadmissible, implying, as they would do, the division of that which, by hypothesis, and by the name founded thereon, is recognized as indivisible. The rules here laid down are well exemplified in the following table of the molecular formulæ, representing the oxides of nitrogen:—

Molecular formulæ of the Nitroxygen series.

1 mol. of nitrous oxide	= 2 at. of nitrogen	+ 1 at. of oxygen	=	N_2O.
1 mol. of nitric oxide	= 1 at. ,,	+ 1 at. ,,	=	$N\ O$.
1 mol. of nitrous acid	= 2 at. ,,	+ 3 at. ,,	=	N_2O_3.
1 mol. of hyponitric acid	= 1 at. ,,	+ 2 at. ,,	=	$N\ O_2$.
1 mol. of nitric acid	= 2 at. ,,	+ 5 at. ,,	=	N_2O_5.

The value of this conception of the molecular structure of elementary as well as compound gases will become more apparent to us in future stages of our inquiry than it is at present. But the distinctness which the atomic theory lends to our views of material phenomena can be immediately perceived. It affords us a satisfactory explanation of the definiteness and immutability of chemical composition; of the step-by-step gradations, in simple multiple ratios, by which the proportions of such elements as form more than one compound with each other are observed to vary; and it enables us readily to understand the fact that compounds in so many cases become less stable as they increase in complexity; in other words, as their molecules are built up of a larger number of atoms.

The incorporation of these views in the formulæ of simple and compound bodies, and of their reactions, evidently impresses upon our symbolic language a new significance, and adapts it to aid in theoretically interpreting the phenomena which it also depicts and records.

We have only to add on this subject that both the atomic and molecular forms of symbolic expression have their peculiar merit; the former being more succinct, the latter more comprehensive. When very complete and encyclopædic expressions are required, including the proportions, both by volume and by weight, as well of the bodies brought into action as of the resultant products, simple and compound, molecular formulæ are indispensable. When only the relative weights of the bodies in action, and of their products, require representation (as in the majority of practical problems), atomic equations are sufficiently comprehensive, and have the advantage in point of conciseness and simplicity. Many chemists indeed use this latter form only of the symbolic language; let it be our care to master both modes of expression, so that we may be able to employ each in turn for its appropriate purposes.

LECTURE X.

Molecular and atomic constitution of the typical compounds—curious relations, ponderal, numerical, and potential, of the typical elementary atoms—two sorts of chemical value or power, molecule-forming and atom-fixing—unitary standard of atom-fixing power—major and minor equivalent weights—coefficients of atom-fixing power, or quantivalence—comparative quantivalence of the typical elements and their congeners—germ of a natural system of chemical classification—volume-condensing power of atoms, how far proportionate to their quantivalence—alternative standards of quantivalence—chemical value in exchange—exemplification thereof in the syntheses of hydrochloric acid and water, and in the hypothetical syntheses of ammonia and marsh-gas—also in their decomposition by chlorine—in the contrasted action of chlorine and oxygen on hydriodic acid—and in the comparative structure of ammonia and nitrous acid—quantivalential equilibrium of the nitroxygen series—distinction between numerical and potential quantivalence—also between quantivalence and chemism—transitional state of the question—tabular summary.

In the light of our new conceptions concerning the molecules and atoms of which the elementary bodies and their compounds are built up, we reviewed, at our last meeting, the diagrammatic symbols previously employed to represent the volumetric and ponderal composition of hydrochloric acid, water, ammonia, and marsh-gas; and we found that the double squares, expressing the *dilitral* product-volumes of these compounds, are perfectly well adapted to represent for us their respective *free molecules*; while the single squares, which previously served us to denote the *monolitral* unit-volumes of their respective constituents, answer equally well to depict the *combining atoms* of those elementary bodies.

In the following diagram the dilitral product-volumes, or, as we must now say, the *molecules*, of our four typical compounds, are placed in a column by themselves, in contrast with the monolitral unit-volumes, or, in our present view, the *atoms* of the elements which they respectively contain; the arrangement being such that the atoms of chlorine, oxygen, nitrogen, and carbon, occupy the second column of the diagram, with the hydrogen-atoms they respectively take up displayed on the right-hand side, and the resulting compound molecules on the left.

MOLECULAR AND ATOMIC CONSTITUTION OF THE FOUR TYPICAL COMPOUNDS.

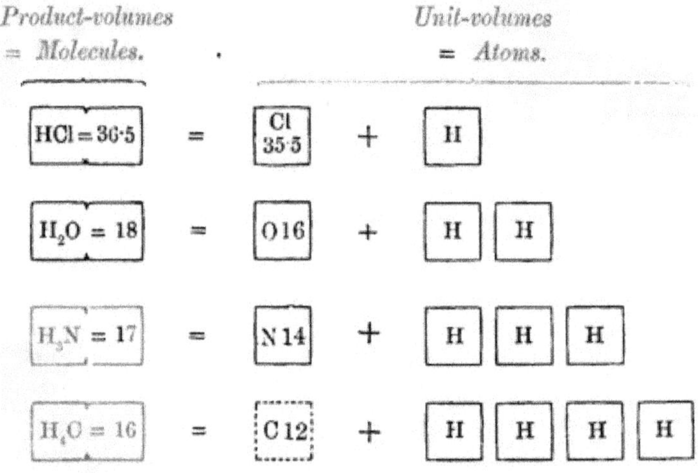

In examining this diagram, we are at once struck with the fact that the four elements, displayed in the second column, stand very differently related, on the one hand, to the volume and weight of the compound molecules they respectively form, and, on the other hand, to the volume and number of the atoms that take part with them in forming those molecules.

Thus, a glance at the left and central columns shows us that, though the volumes of the compound molecules are equal, they contain very unequal weights of the four elements under consideration, viz., 35·5 of chlorine, 16 of oxygen, 14 of nitrogen, and 12 of carbon respectively.

Again, looking to the number of hydrogen-atoms depicted on the right-hand side, we see that the atoms of the four central elements, chlorine, oxygen, nitrogen, and carbon, stand related respectively to 1, 2, 3, and 4 hydrogen-atoms.

It is a curious circumstance, and one which the diagram almost forces us to notice, that the heaviest of the four central atoms (Cl = 35·5) is precisely the one which engages the smallest number of hydrogen-atoms, viz., only one; while the other three (O = 16, N = 14, and C = 12), as they grow successively lighter, engage increasing numbers of hydrogen-atoms, viz., 2, 3, and 4 atoms respectively.

In other words, it takes the whole atom-power of chlorine, 35·5, to engage 1 atom of hydrogen; whereas, the atom-power of oxygen, 16, suffices to engage 2 hydrogen-atoms; and the atom-power of nitrogen and carbon suffice, respectively, to engage 3 and 4 hydrogen-atoms.

It is impossible to overlook the singular numerical relations which these unequal atom-engaging powers bring about in the product-volumes, or molecules, depicted on the left side of the diagram. While carbon, nitrogen, and oxygen have, as their atom weights, 12, 14, and 16, respectively, the molecules they form, in combining with hydrogen, weigh 16, 17, and 18 respectively; the molecular weights advancing by 1 only, while the atomic weights advance by 2, at each grade. This is made still more remarkable by the fact that from 18, the weight of the water-gas molecule, to 36·5, the weight of the hydrochloric acid molecule, the advance is by a sudden spring to *about* double the first-named quantity; 18 × 2 being 36, while the actual weight of the hydrochloric acid molecule is 36·5. Another element (*fluorine*), to which our attention has not been directed, fills up, in an equally interesting manner, an intermediate link in this numerical chain, as we shall hereafter learn. These curious relations are entirely unexplained, though they have latterly attracted much attention. They do not, however, belong to our present inquiry; from which, it must be owned, we have digressed for a moment to bestow on them this passing notice.

Returning to our immediate study, we observe that the table places before us the four centrally-disposed elements, in two perfectly distinct chemical relations; the first more especially *volumetric* and *molecular*, the second essentially *numerical* and *atomic.* Hence, two parallel series of minimum-weights; one representing the minimum quantity of each element requisite to take part in the formation of a compound *molecule*; the other corresponding to the minimum quantity of each element which is adequate to engage or fix one standard *atom*.

As the standard atom, whereby to measure this atom-fixing power of the elements, the table presents us with hydrogen (H = 1), and this unit we gladly adopt: first, because hydrogen is also our unitary standard for relative volume-weight or specific gravity; and for concrete volume-weight (*i.e.*, specific gravity read in criths); secondly, because hydrogen is, as we shall presently see, particularly well suited to serve as our unit for this additional purpose.

Working out the comparison of these two sorts of chemical value, the *molecule-forming* and the *atom-fixing*, for hydrogen and the four other elementary bodies under review, we obtain the three following columns of figures:—

CHEMICAL VALUES, molecule-forming, and atom-fixing, of the standard element, hydrogen, and of the four typical elements; with the ratios of those values.

The four Typical Elements, preceded by the Standard Element, Hydrogen.		Minimum weights thereof requisite		Ratios of the numbers in Columns 3 and 4.
Their names.	Their literal symbols.	To take part in the formation of a molecule.	To engage one standard atom.	
1.	2.	3.	4.	5.
Hydrogen .	H	1	1	$\frac{1}{1} = 1$
Chlorine . .	Cl	35·5	35·5	$\frac{35\cdot5}{35\cdot5} = 1$
Oxygen . .	O	16	8	$\frac{16}{8} = 2$
Nitrogen . .	N	14	4·66	$\frac{14}{4\cdot66} = 3$
Carbon . .	C	12	3	$\frac{12}{3} = 4$

In this table, it will be observed, column 3 represents the *molecule-forming equivalents* of the elements, or the proportions by weight in which they can replace each other in contributing to the construction of a *molecule;* while column 4 sets forth the *atom-fixing equivalents* of the elements, or the proportions in which they can replace each other in fixing a standard *atom*. Column 5 shows the *ratios* of the molecule-forming to the atom-fixing weights; and the usefulness of these ratios we will now proceed to consider.

Taking, for example, the last element in the table, carbon, we see that its molecule-forming minimum-weight is 12, while its standard-atom-fixing minimum-weight is 3; and, on referring to the diagrammatic formula of marsh-gas (p. 106) we find that, in reality, 12 parts by weight of carbon enter into the construction of the marsh-gas molecule, while these 12 parts fix, in that molecule, 4 atoms of hydrogen; each of which, therefore, is fixed by $\frac{12}{4} = 3$ parts by weight of carbon.

So likewise, of nitrogen (the element next above carbon in the table), we see that 14 is the molecule-forming minimum, and $\frac{14}{3} = 4{\cdot}66$ the atom-fixing minimum; of oxygen, these two minima are respectively 16 and $\frac{16}{2} = 8$; while, for chlorine, which fixes only 1 standard atom, the two minima of course coincide.

As for hydrogen, it stands apart, at the head of each column, the unitary standard of atom-fixing power.

From these considerations it is clear that we might attach to each element *two* representative or equivalent numbers; one expressing its minimum-weight relatively to the formation of a *molecule*, the other its minimum-weight relatively to the fixation of an *atom;* and we might distinguish these as its *major* and *minor* equivalent weights, or by some other distinctive designations.

It would, however, be obviously inconvenient thus to have in use two sets of *minima* weights, or equivalents. Such a duplicate system of notation would encumber the memory, and greatly impair the succinctness of our symbolic short-hand. In order to avoid these evils it is desirable to include, for each element, the two separate weights in a single concise expression.

For this purpose it suffices to attach, to each of the molecule-forming minimum-weights, given in the third column of the table, a *coefficient of atom-fixing power;* that is to say, a sign expressing how many standard atoms its said weight is adequate to satisfy. This is readily done by aid of the ratios set forth in column 5 of the table. These ratios are, in fact, the coefficients in question; and by writing them (in Roman numerals for distinctness' sake) over against the molecule-forming minimum-weights, in the ordinary place of exponents, we learn at a glance the number of standard atoms which the said weights can respectively fix.

Accordingly, the molecule-forming minimum of chlorine = 35·5 is written $35{\cdot}5^{\text{I}}$; the corresponding minimum of oxygen = 16 is written 16^{II}; that of nitrogen = 14 is written 14^{III}; and that of carbon = 12 is written 12^{IV}. Or, for still greater brevity, as the signs Cl, O, N, and C, are already associated in our memory with their respective molecule-forming weights, we may attach the coefficients directly to these symbols, writing them respectively:—

$$Cl^{\text{I}},\ O^{\text{II}},\ N^{\text{III}},\ \text{and}\ C^{\text{IV}}.$$

Dashes, one, two, three, or four in number, are commonly employed by chemists instead of numerals to express these coefficients; and they answer the purpose equally well for the lower expressions. But, for numbers higher than three, the numerical expressions are preferable to the dashes, as being easier than they, both to write and read; and, to preserve uniformity throughout, I prefer to use Roman numerals for these signs in all cases.

We are in want of a good appellation to denote this atom-fixing power of the elements. The vague and rather barbarous expression, *atomicity*, has drifted into use for this purpose; and the elements have been called *mon*atomic, *di*atomic, *tri*atomic, and *tetra*tomic, accordingly as their respective molecule-forming minimum-weights are capable of saturating 1, 2, 3, or 4 standard atoms.

These expressions are faulty, because they are open to misinterpretation, as if intended to denote the atomic structure of the respective elementary molecules themselves; a sort of confusion, the possibility of which should always be sedulously avoided in scientific nomenclature.

We shall escape this evil by substituting the expression *quantivalence* for *atomicity;* and designating the elements *uni*-valent, *bi*valent, *tri*valent, and *quadri*valent, according to their respective atom-fixing values.

As to their molecule-forming values, these may be indifferently termed their atom-weights, or combining numbers; which, it will be remembered, correspond, for the volatile elements, with a few exceptions only, to their respective gas-volume-weights, or specific gravities relatively to hydrogen.

However denoted, the two sorts of chemical value remain in themselves most clearly distinguished; and the importance of keeping this distinction in mind cannot be overrated. By the difference of their *quantivalent* powers the four elements under review are impressed, each with a strongly-marked character of its own; and we shall presently see that each of the four stands in this respect at the head of a group of congeners, endowed with like atomic quantivalence; so that we have here the first germ of a natural *classification* of chemical elements, based on experiment, and conformable with truth.

Thus, for example, to univalent chlorine (Cl^{I}) correspond bromine and iodine, both likewise univalent, and written accordingly Br^{I} and I^{I}.

So, again, to bivalent oxygen (O^{II}) correspond bivalent sulphur and selenium—S^{II} and Se^{II} respectively.

Trivalent nitrogen (N^{III}) has for its congeners trivalent phosphorus and arsenic, P^{III} and As^{III}.

Lastly, quadrivalent carbon (C^{IV}) is the typical head of the group which comprises quadrivalent silicon and titanium, written Si^{IV} and Ti^{IV}.

It should not be overlooked that, with the varied atom-fixing or quantivalent powers are coupled, in the case of the typical elements, and of their respective combinations with hydrogen, proportionate *volume-condensing* powers; these two powers (as we remember) going together, and increasing *pari passu*. Thus, the four atoms of hydrogen fixed by quadrivalent carbon in marsh-gas are not only four times as numerous, but also four times as compactly disposed, as the one hydrogen-atom fixed by univalent chlorine in hydrochloric acid gas. To this rule, however, phosphorus and arsenic are, to a certain extent, exceptions. These trivalent bodies do, indeed, fix as many hydrogen-atoms as their trivalent architype, nitrogen; and they fix them, moreover, within the same dilitral product-volume. Yet, remembering as we do that the normal nitrogen-atom occupies twice as much space as the exceptional atoms of phosphorus and arsenic respectively, we are fain to admit that the space left for hydrogen-atoms in the respective product-volumes of phosphoretted and arsenetted hydrogen is greater, by exactly half a volume, than the space left for hydrogen-atoms in the product-volume of ammonia. In other words, the total condensation in the case of ammonia is represented by the ratio $\frac{2}{4} = \frac{1}{2}$; whereas the total condensation, in the cases of phosphoretted and arsenetted hydrogen, is only in the ratio $\frac{2}{3\cdot5} = \frac{1}{1\cdot75}$. As for carbon, its free vapour having never been obtained, we have no means of determining the space filled by its atom in gaseous compounds. It is only therefore provisionally, and on analogical, not experimental, grounds, that we place it volumetrically on a par with the other elements under consideration. Hence the dotted lines with which (as before) the carbon-atom is represented in our diagram.

How far the coincidence between the atom-fixing and volume-condensing powers of the typical elements observed in their combinations with hydrogen, may extend to their combinations with other bodies, and how far such coincidence may be traceable in the combining relations of the elements generally, these are problems as yet unsolved.

Reverting to the simple consideration of quantivalence itself, we of course understand that the expressions uni, bi, tri, and quadrivalent would have to be modified, if the atom selected as a unitary standard of quantivalence were itself of other than univalent power. For example, were the bivalent oxygen-atom selected, as the standard or unit of quantivalence, hydrogen, having only half that value, would have to be called *semivalent;* while nitrogen, having once and a half that value, would be *sesquivalent*; and carbon, having twice that value, would be bivalent. Thus, though the degrees of quantivalence would be quite as correctly expressed in terms of a bivalent as of a univalent standard, two expressions out of the four would be *fractional* —a manifest inconvenience. This inconvenience would become increased—the fractional expressions would become more numerous, were a trivalent or quadrivalent atom, such as the atom of nitrogen, or that of carbon, adopted as the standard of quantivalence. If, therefore, we had not already freely chosen univalent hydrogen as our standard of quantivalence, we should have been constrained, by the exhaustive method, to its adoption, as the simplest and best for the purpose

The unequal *molecule-forming* powers of the elementary bodies, represented by the different weights of their atoms, and their unequal *atom-fixing* powers, represented by their dissimilar coefficients of quantivalence, show us that each of these bodies possesses what may be termed its *specific chemical value in exchange.*

Thus, with respect to the power of forming a molecule, we know already, and are again reminded by looking at our last diagram, that 12 parts by weight of carbon are "worth" as much for this purpose as 14 parts of nitrogen, 16 of oxygen, and no less than 35·5 of chlorine.

So again, with reference to the power of fixing a standard atom, the elements comprised in each of the four groups whereof chlorine, oxygen, nitrogen, and carbon are the respective types, possess for this purpose *chemical value in exchange*, varying by gradations, as we have just seen, from 1, for the first group, through 2 and 3 for the second and third, up to 4 for the last.

In other words, and to borrow a financial mode of expression, one atom of any element in group 4 (the quadrivalent group) is *exchangeable at par* for *four* atoms of any element in group 1 (the univalent group), and for two atoms of any element in group 2 (the bivalent group). So also, as to the reciprocal exchangeability at par of the atoms of elements comprised respectively in the trivalent and quadrivalent groups, the simplest way to represent this equation, without resorting to fractional expressions, is to describe *three* atoms of any of the quadrivalent elements as "worth," for atom-fixing purposes, *four* atoms of either of the trivalent bodies.

From this conception of "value in exchange" we readily gather that the atomic relations which we designate quantivalence imply not only atom-*fixing*, but also atom-*displacing* power, so that, in learning how many standard units of quantivalence any given elementary atom can attract and retain within a compound molecule, we learn also how many it can remove therefrom, when it is employed as a decomposing agent, under conditions enabling it to eliminate, partly or wholly, one or more of the constituents thereof.

The differences between the elements, as to their chemical *value in exchange*, for molecule-forming and atom-fixing (or freeing) purposes, are well illustrated in the simple equations which represent the syntheses of hydrochloric acid and water respectively.

We referred to these equations when speaking of the molecular structure of the typical elementary gases; and we then pointed out that, when two such gases combine directly to form a compound gas, there is a reciprocal exchange of atoms between their diatomic molecules; which exchange, we may now add, bespeaks, by its conditions, their relative molecule-forming and atom-fixing powers.

Contrast, for example, the equation representing the synthesis of hydrogen and chlorine (both univalent elements) to form hydrochloric acid, with the equation representing the synthesis of hydrogen and oxygen (the latter a bivalent element) to form water.

In the first case we have—

Synthesis of Hydrochloric Acid.

Constituent Gases. *Hydrochloric Acid Gas produced.*

$$\boxed{HH} + \boxed{Cl^{I}Cl^{I}} = \boxed{HCl^{I}} + \boxed{HCl^{I}}$$

$$\text{Or } HH + Cl^{I}Cl^{I} = 2\ HCl^{I}$$

Each atom of univalent chlorine here passing in exchange, at par, for one atom of univalent hydrogen; so that we may regard hydrochloric acid (HCl) as a molecule of hydrogen (HH) in which one atom of hydrogen (H) is replaced by one atom of chlorine (Cl), (or *vice versâ*).

In the second case we have a very different expression; for here bivalent oxygen comes into play, and we accordingly obtain the equation:—

Synthesis of Water.

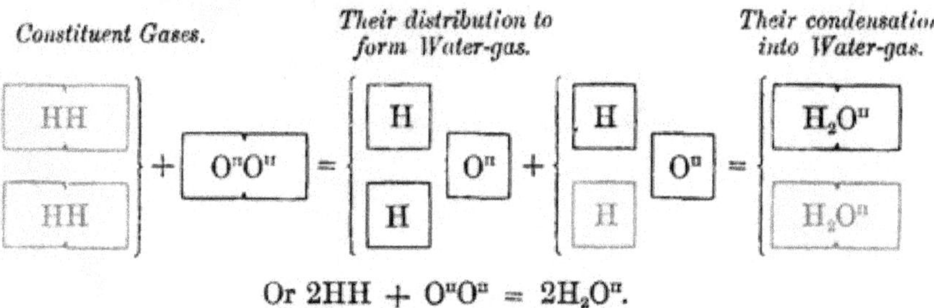

$$\text{Or } 2HH + O^{II}O^{II} = 2H_2O^{II}.$$

In the water-gas molecule we see bivalent oxygen fixing twice as many atoms of hydrogen as univalent chlorine fixes in hydrochloric acid; and the equation shows us that, in the act of combination, two standard univalent hydrogen-atoms are, so to speak, *bartered*, or given in exchange at par, for one atom of divalent oxygen.

In the two following diagrams we have represented, in like manner, for the further illustration of our subject, the (as yet theoretical) syntheses of ammonia and marsh-gas; employing as before, it will be observed, dotted squares to signify the hypothetical character of carbon vapour. Though these syntheses have not yet been directly accomplished, they obviously imply the play of elements here depicted; and we thus learn that, just as in the formation of hydrochloric acid gas and water-gas, respectively, univalent chlorine replaces 1 atom, and divalent oxygen 2 atoms, of hydrogen, just so, in the production of ammonia and marsh-gas, the trivalent nitrogen-atom, and the quadrivalent carbon-atom, respectively replace 3 and 4 atoms of the same unitary standard.

Synthesis (hypothetical) of Ammonia.

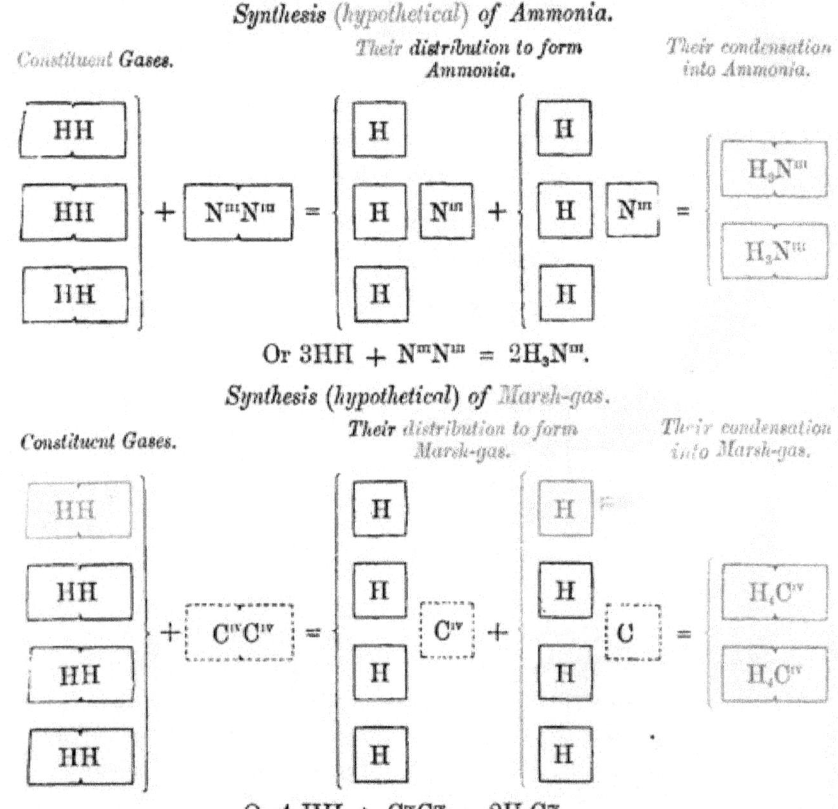

Or $3HH + N^{III}N^{III} = 2H_3N^{III}$.

Synthesis (hypothetical) of Marsh-gas.

Or $4\ HH + C^{IV}C^{IV} = 2H_4C^{IV}$.

Nor is it only by their *synthetic* equations that our four typical compounds illustrate the dissimilar quantivalence of the elementary atoms; we shall find their equations of decomposition equally instructive on this head.

We remember, for example, employing chlorine to expel oxygen from water, nitrogen from ammonia, and carbon from marsh-gas; and, keeping in view the quantivalence of these bodies respectively, we may be sure that, of the univalent chlorine, 2 atoms are requisite to expel from the water-gas molecule its atom of divalent oxygen, 3 to remove from the ammonia molecule its atom of trivalent nitrogen, and 4 to replace in the marsh-gas molecule its atom of quadrivalent carbon.

Experiments, which we have already made together, have established these relative values as indisputable facts; and they are compendiously displayed in the following symbolic equations.

Decomposition of Water by Chlorine.

1. *Atomic expression.*

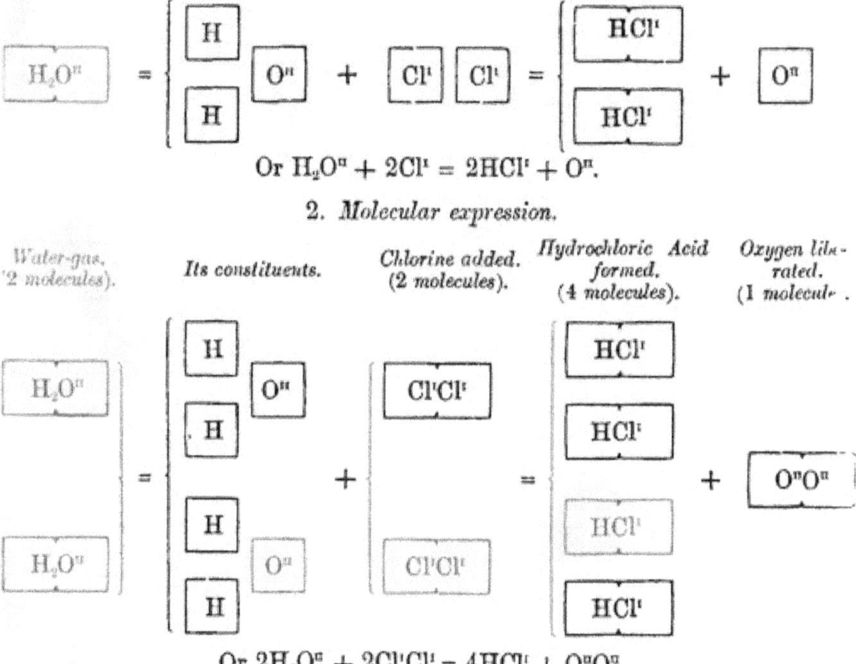

Or $H_2O^{ii} + 2Cl^{i} = 2HCl^{i} + O^{ii}$.

2. *Molecular expression.*

Or $2H_2O^{ii} + 2Cl^{i}Cl^{i} = 4HCl^{i} + O^{ii}O^{ii}$.

Decomposition of Ammonia by Chlorine.

1. *Atomic Expression.*

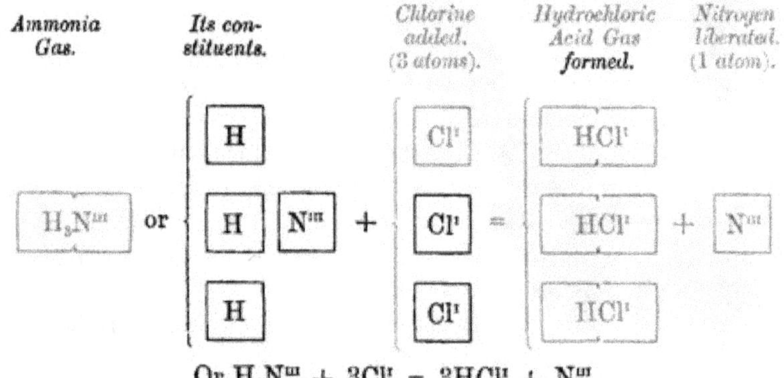

Or $H_3N^{iii} + 3Cl^{i} = 3HCl^{i} + N^{iii}$.

2. *Molecular expression.*

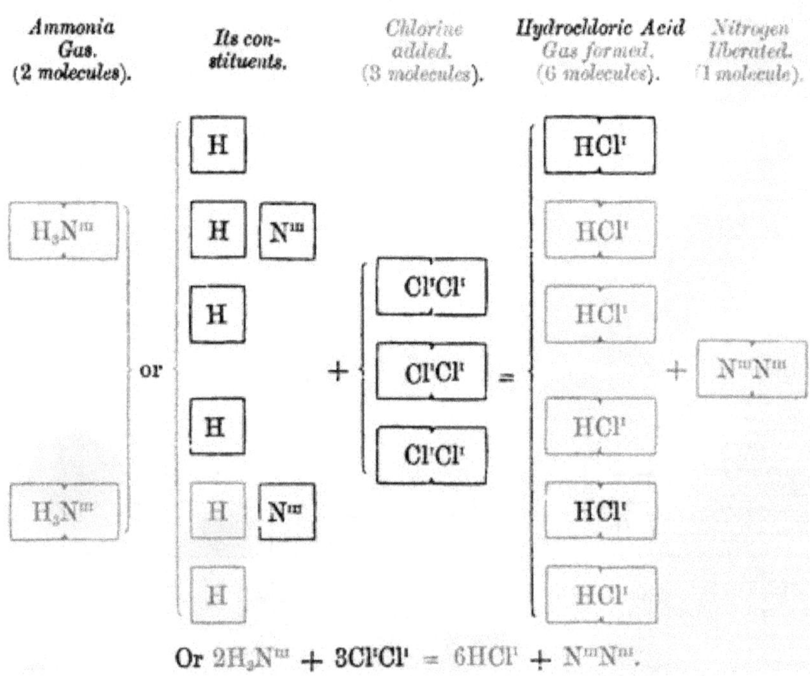

Or $2H_3N^{iii} + 3Cl^{i}Cl^{i} = 6HCl^{i} + N^{iii}N^{iii}$.

Decomposition of Marsh-gas by Chlorine.

1. *Atomic expression.*

Marsh-gas.
Its constituents.
Chlorine added. (4 molecules).
Hydrochloric Acid Gas formed.
Carbon liberated. (1 atom).
H_4C^{iv} or
H H H H C^{iv} +
Cl^i Cl^i Cl^i Cl^i =
HCl^i HCl^i HCl^i HCl^i + C^{iv}

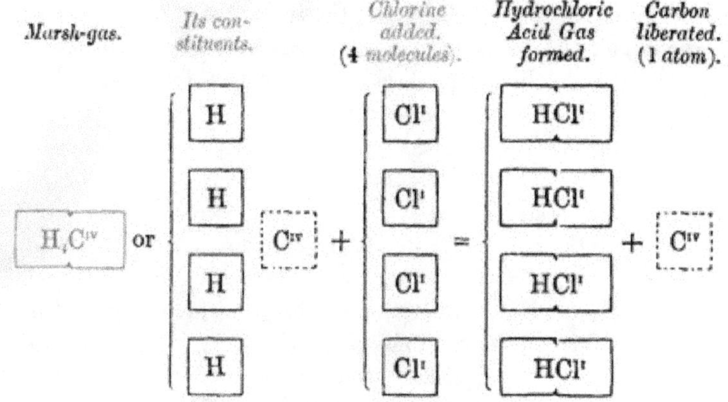

Or $H_4C^{iv} + 4Cl^i = 4HCl^i + C^{iv}$.

2. *Molecular expression.*

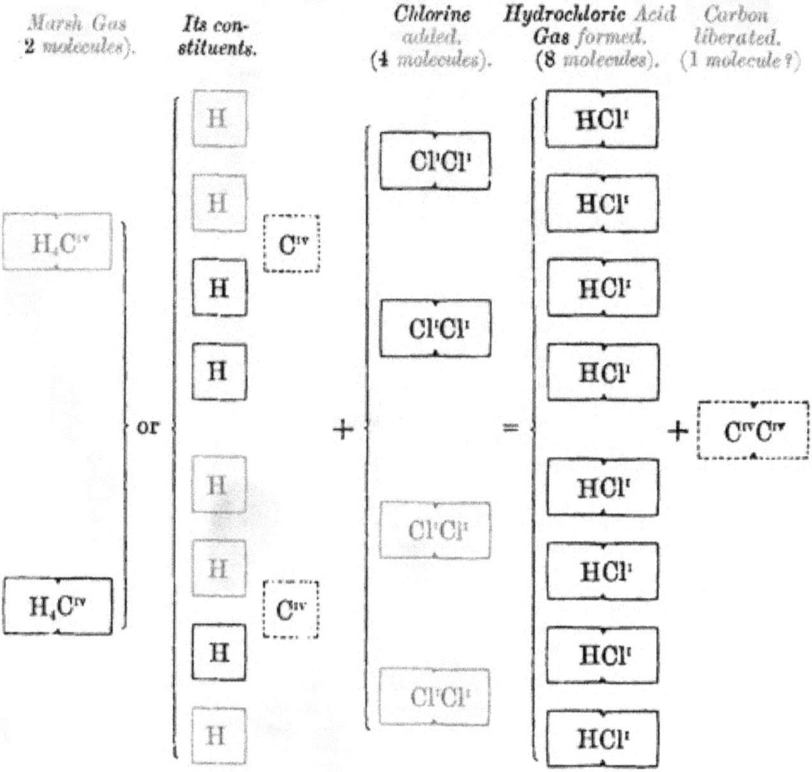

Marsh Gas (2 molecules).
Its constituents.
Chlorine added. (4 molecules).
Hydrochloric Acid Gas formed. (8 molecules).
Carbon liberated. (1 molecule?)
H_4C^{iv}
H_4C^{iv}
or
H H C^{iv} H H
H H C^{iv} H H
+
Cl^iCl^i Cl^iCl^i Cl^iCl^i Cl^iCl^i
=
HCl^i HCl^i HCl^i HCl^i HCl^i HCl^i HCl^i HCl^i
+ $C^{iv}C^{iv}$

Or $2H_4C^{iv} + 4Cl^iCl^i = 8HCl^i + C^{iv}C^{iv}$.

Each of the foregoing equations, it will be observed, is given in two forms; one atomic, the other molecular. The former is the simpler, and, to the beginner, the more intelligible mode of expression; but, as it frequently depicts free gases as if composed of isolated atoms, it is to that extent incorrect. The molecular expressions represent the free gases as of their true diatomic structure; and, accordingly, though more cumbrous, they are more strictly accurate symbolizations. By the instructed eye these latter will therefore be preferred; and by contrasting the two forms, in the few previous cases, we obtain an opportunity of familiarizing ourselves with both.

Another excellent illustration of this subject is supplied by the contrasted behaviour of chlorine and oxygen, when employed successively to expel iodine from hydriodic acid. Iodine, as we remember, is a congener of chlorine, and, like it, univalent; whereas oxygen is our typical bivalent. Oxygen ought, therefore, to display twice the atom-expelling power of chlorine; wherefore, if chlorine, as a univalent body, expels from hydriodic acid 1 atom of iodine (also univalent), oxygen ought evidently to expel therefrom two. In other words, for 1 molecule of hydriodic acid decomposed by a chlorine-atom, 2 should be decomposed by an oxygen-atom. This ratio of quantivalence, or atomic value in exchange, is actually established by experiment. For any given weight of hydriodic acid attacked, and any given number of iodine-atoms expelled therefrom, by univalent chlorine, twice the weight and double the number are attacked and expelled by bivalent oxygen. For the display of these results in the diagrams subjoined, we have selected the more precise molecular form of expression; which, by aid of the above models, you can readily transform, if so minded, into the simpler atomic equations.

In each case we have, it will be observed, as in the previous diagrams, placed first the symbol of the compound to be decomposed, next this the element added for its decomposition, and afterwards, in due succession, the products, elementary and compound, of the reaction; interposing, when necessary, a sketch of the redistribution of the atoms present, to form the new products.

Decomposition of Hydriodic Acid by Chlorine.

Hydriodic Acid Gas. (2 *molecules*).		*Chlorine added.* (1 *molecule*).		*Hydrochloric Acid Gas formed without change of volume.* (2 *molecules*).		*Iodine liberated.* (1 *molecule*).
$HI^{\prime}$ / $HI^{\prime}$	+	$Cl^{\prime}Cl^{\prime}$	=	$HCl^{\prime}$ / $HCl^{\prime}$	+	$I^{\prime}I^{\prime}$

$$\text{Or } 2HI^{\prime} + Cl^{\prime}Cl^{\prime} = 2HCl^{\prime} + I^{\prime}I^{\prime}.$$

Decomposition of Hydriodic Acid by Oxygen.

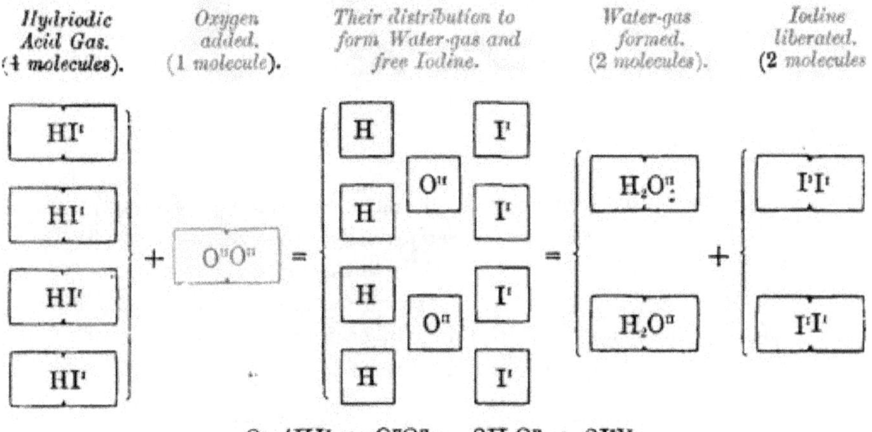

$$\text{Or } 4HI^{\prime} + O^{\prime\prime}O^{\prime\prime} = 2H_2O^{\prime\prime} + 2I^{\prime}I^{\prime}.$$

Once more, if we compare the quantivalent powers of univalent hydrogen and bivalent oxygen, as manifested in the quantity of trivalent nitrogen which they respectively engage in forming ammonia and nitrous acid, we obtain another exact and beautiful confirmation of the law of quantivalence. This is displayed in the following diagram, in which, it will be observed, the molecule of nitrous acid gas is represented in dotted outline, our usual sign of doubt; the vapour-density of this body (like that of nitric acid) remaining still to be determined:—

Comparison of the Structure of Ammonia and Nitrous Acid.

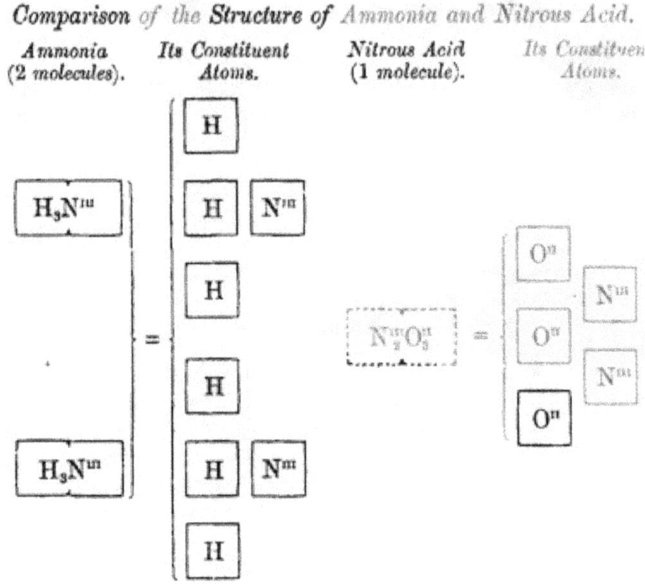

This diagram clearly sets before us, on one side (in ammonia), 2 atoms of trivalent nitrogen reciprocally fixing, and fixed by, 6 atoms of univalent hydrogen; while to the right, 2 atoms of trivalent nitrogen, in nitrous acid, fix, and are fixed by, 3 atoms only of bivalent oxygen. Multiplying, in each case, the numbers of atoms engaged, by their coefficients of quantivalence, we have on both sides alike, 6 units of quantivalence balancing 6.

Thus for ammonia we have $N^{\text{III}\times 2=\text{VI}}$ balancing $H^{\text{I}\times 6=\text{VI}}$,

While for nitrous acid we have $N^{\text{III}\times 2=\text{VI}}$ balancing $O^{\text{II}\times 3=\text{VI}}$.

It is by no means, however, to be supposed that the atom-fixing or atom-expelling power of the elementary bodies, or, as we have agreed to term it, their quantivalence, is under all circumstances, and in all the compounds which they form, invariably exerted to its fullest extent. Were the chemical combination of elementary atoms governed in this respect by a fixed and unbending law, it would evidently be impossible for any two given elements to unite in any other proportion than that implying their complete reciprocal satisfaction. Trivalent nitrogen and divalent oxygen, for example, could, under such a law, unite only in the proportion of two atoms of the former body to three of the latter to form the compound $N^{iii}_2O^{ii}_3$, which we have just been comparing with ammonia; whereas we already know that these two elements form, by their varied combinations, no less than 5 different compounds, which we have studied under the name of the *Nitroxygen series* (compare p. 183). Of this series one member only, viz., $N^{iii}_2O^{ii}_3$, enjoys what may be termed a perfect quantivalential equilibrium; that is to say, a complete reciprocal satisfaction of their quantivalence by the elements present.

It has been recently pointed out (comp. Preface, p. vi.) that, while this perfectly balanced, or self-satisfied body, $N^{iii}_2O^{ii}_3$, is the central body, or pivot, of the nitroxygen series, this series itself, taken collectively, presents an equally perfect equipoise, or quantivalential symmetry; its four remaining members being disposed in pairs, one pair on either side of the centre,—the members of the left-hand pair containing an excess of N^{iii} quantivalence unbalanced by O^{ii}; while those of the right-hand pair, on the contrary, possess an excess of O^{ii} quantivalence uncompensated by N^{iii}.

The interest of this observation (now, I believe, first published) lies mainly in the curious fact, that these opposite excesses exactly match, or, so to speak, inversely reflect each other; a fact which its observer has constructed the following diagram to display.

Nitroxygen Compounds.
Their Quantivalential Composition.

Hypo-oxygenic.		*Equiquantic.*	*Hypo-nitrogenic.*	
Nitrous Oxide.	*Nitric Oxide.*	*Nitrous Acid.*	*Hyponitric Acid.*	*Nitric Acid.*
$N^{iii}_2O^{ii}$	$N^{iii}O^{ii}$	$N^{iii}_2O^{ii}_3$	$N^{iii}O^{ii}_2$	$N^{iii}_2O^{ii}_5$
Their Quantiquivalential Proportions.				
6 to 2	3 to 2	6 to 6	3 to 4	6 to 10
Their Quantivalential Deficiencies.				
4	1	0	1	4

The foregoing diagram, it will be observed, is so arranged as to exhibit prominently, in the centre of the series, that one of the five nitroxygen compounds in which the opposite quantivalential or atom-fixing powers are equally balanced; a condition which, to avoid periphrasis, we will agree, if you please, to term *equiquantic*. From this central compound the others are shown falling away on either side, to represent their progressive *declension* from equiquanticity. It will be observed that each *case* or decline (*case* is from the Latin *casus*, a fall) that takes place on the side of deficiency in $N^{\prime\prime\prime}$ quantivalence, is matched by a corresponding case, or decline, taking place on the side of deficiency in $O^{\prime\prime}$ quantivalence; so that the series is perfectly symmetrical.

It will be also observed that the deficiency of quantivalence in each of the terminal compounds of the series (nitrous oxide and nitric acid) is = 4; which deficiency diminishes to = 1 only, for each of the two compounds (nitric oxide and hyponitric acid) that intervene between the terminals and the central or equiquantic compound (nitrous acid).

There is certainly much beauty in the perfect symmetry and absolute equipoise of this natural chemical group; with its two equal wings spreading, so to speak, from the central body, each, in respect of its quantivalential relations, the very counterpart of the other.

Similar deficiencies of quantivalence occur, of necessity, in every other series of binary compounds exemplifying, like the nitroxygen series, the law of multiple proportions. Such deviations are also frequently observed in compounds of higher complexity, belonging to series of greater extent. The conditions of their occurrence, and the characters they impress upon the compounds thus unequally composed, are subjects for experimental study, and philosophical meditation, of the very deepest interest. We shall hereafter find that bodies which are of *numerically* equiquantic structure are not always, on that account, the most stable; while, on the other hand, *numerical* deficiencies of quantivalence often coexist with what may be termed *potential* equiquanticity. For example, in the nitroxygen series itself, the central equiquantic body, $N^{\prime\prime\prime}_2O^{\prime\prime}_3$, is not, as might be expected, the most stable compound of the series; on the contrary, it breaks up readily under various influences; whereas the terminal member of the left wing, $N_2^{'''}O''$, notwithstanding its hypo-oxygenic structure, possesses remarkable stability. Again, $N_2^{'''}O_5''$, the terminal member of the opposite or hypo-nitrogenic wing, though it is numerically on a par, in point of quantivalential deficiency, with $N_2^{'''}O''$ (the deficit being represented by 4 in both cases), instead of possessing the stability of that body, cannot be vaporized without breaking up into compounds of a lower order. These apparently anomalous facts depend upon conditions into the consideration of which we cannot here enter. At present it must suffice us to remark, in passing, that, *cæteris paribus*, compounds in which the quantivalents are numerically balanced are more readily formed, and of more stable character, than those in which either constituent is in quantivalential excess.

At this point a few words may be usefully interposed to guard the student from misconceiving the true import of the term *quantivalence*. This word is employed to designate the particular numerical atom-compensating power inherent in each of the elements, and this power of theirs must by no means be confounded with the specific *intensity* of their respective chemical activities. Thus, for example, nitrogen, phosphorus, and arsenic are all of them *tri*valent bodies; but this equality of their atom-compensating values does not imply that they are all endowed with equal *avidity* for this or that element—say oxygen or hydrogen, for example. Because one *quadri*valent carbon atom can fix four atoms of hydrogen, of which one only is fixed by *uni*valent chlorine, we are not, therefore, to suppose that the attraction of carbon for hydrogen is necessarily more intense than that of chlorine, in the ratio $\frac{1}{4}$. On the contrary, in this case, the element which seizes the lesser proportion of hydrogen, seizes it with by far the greater degree of energy. For another and equally cogent illustration of this truth we may refer to our very first experiment. In this, as you remember, we saw potassium (a univalent body, as we shall presently learn), take oxygen from its combination with hydrogen (also univalent) in water; equalit of quantivalence coexisting, in this case, with a strongly-marked difference in the degree or *intensity* of that peculiar endowment which is commonly denominated *chemical attraction*, but to which the shorter and more abstract appellation *Chemism* may be more safely and appropriately assigned. The term *chemism* is derived from the same root which also supplies *chemistry* itself with its name (comp. p. 2); and *chemism* seems a fitting designation for that peculiar power, or endowment, which constitutes the special subject of the science analogously named. The phenomena of *chemism*, and the conditions of its varying intensity, have hitherto but cursorily engaged our attention, and rather as incidental to experiments made with other subjects of study in view, than as themselves claiming immediate notice. We shall have to examine those phenomena in their turn hereafter; and we shall then learn, amongst other things, how much the relative *chemism* of bodies varies, so far at least as its results are concerned, with the varying physical conditions under which it is called into play. We merely refer to this subject here, that we may note in passing, the strong lines of demarcation which divide the *chemism* of the elements from their *molecule-forming* power, on the one hand, and, on the other, from their *atomic quantivalence*.

With reference to this latter property of the elementary bodies, and to the relations which subsist between its laws and those of combination in multiple proportions, much as these subjects tempt us to further investigation, this must, for the present, stand adjourned. We must, indeed, become acquainted with a greater variety of chemical compounds before we can further pursue with advantage this branch of our inquiry. The scanty materials as yet at our disposal do not enable us fully to appreciate the assistance afforded to modern chemical research by the development of the laws which govern the unequal molecule-forming and atom-fixing powers of the elementary bodies. In these respects modern chemistry itself is, indeed, in a state of rapid transition; each year—not to say every day—adding something to our knowledge of the atomic and molecular endowments of the elementary bodies and their compounds; of the analogies which tend to enrich with new members the groups they respectively represent; and of the exceptions—apparent or real—which, for the present, obscure the boundaries of those groups. For the perfect elucidation of these deeply-interesting problems we must await the "long results of time"—and must all be contented, whether teaching or taught, humbly to labour and learn, side by side, in the school of the great Teacher—NATURE.

The following Table is an extension of the one above given, comprising the four typical elements only.

TABLE of the MINIMUM MOLECULE-FORMING and MOLECULAR WEIGHTS of SEVERAL ELEMENTS, with their coefficients of quantivalence.

Names of the typical elements and their congeners; preceded by that of the standard element. 1.	Minimum molecule-forming weights; also called *combining numbers*, and *atom-weights*; with their atomic symbols and coefficients of quantivalence. 2.	Molecular weights and their symbols; showing the atomic structure of the elementary molecules in their uncombined state. 3.
Hydrogen .	$H = 1$	$HH = 2$
Chlorine . .	$Cl^{I} = 35{\cdot}5$	$Cl^{I}Cl^{I} = 71$
Bromine . .	$Br^{I} = 80$	$Br^{I}Br^{I} = 160$
Iodine . .	$I^{I} = 127$	$I^{I}I^{I} = 254$
Oxygen . .	$O^{II} = 16$	$O^{II}O^{II} = 32$
Sulphur . .	$S^{II} = 32$	$S^{II}S^{II} = 64$
Selenium .	$Se^{II} = 79$	$Se^{II}Se^{II} = 158$
Nitrogen . .	$N^{III} = 14$	$N^{III}N^{III} = 28$
Phosphorus .	$P^{III} = 31$	$P^{III}P^{III}P^{III}P^{III} = 124$
Arsenic . .	$As^{III} = 75$	$As^{III}As^{III}As^{III}As^{III} = 300$
Carbon . .	$C^{IV} = 12$	$C^{IV}C^{IV}$? $= 24$
Silicon . .	$Si^{IV} = 28{\cdot}5$	$Si^{IV}Si^{IV}$? $= 57$
Titanium . .	$Ti^{IV} = 50$	$Ti^{IV}Ti^{IV}$? $= 100$

In this Table all the simple bodies we have had under our notice are set forth in the first column, with their minimum molecule-forming weights, or atom-weights, set against them in the second column; while, in the third column, these atomic expressions and weights are converted by appropriate modifications into *molecular* symbols. The coefficients of quantivalence are given in both columns; and hydrogen, as before, is placed alone at the head of the Table, as the unitary standard to which all the other bodies are referred. These latter, it will be observed, are disposed in the four typical groups, now (I trust) so familiar to us all.

LECTURE XI.

Principles established for 13 elements, how far applicable to the remaining 48—method of determining the atom-weights of elements, neither volatile themselves, nor capable of forming volatile compounds with hydrogen—value of volatile chlorine-compounds for this purpose—also of volatile oxygen and nitrogen compounds when such exist—illustrations of the principle supplied by such as are volatile of the chlorine, oxygen, and nitrogen compounds of hydrogen, sodium, and potassium—atom-weights and quantivalence of sodium and potassium as determined by the balance—prospect of volumetric corroboration of these results—further illustrations—mercury—bismuth—tin—their vapour-densities—ponderal composition of their respective chlorides—deductions therefrom of their respective atom-weights and coefficients of quantivalence—universal applicability of this method—its liability to error in one respect—further data requisite to obviate this—such data how to be obtained—curious exceptional relation of the atomic and molecular weights of mercury—as also of cadmium—contrast presented by these metals in this respect to phosphorus and arsenic—contrasted symbolization of these bodies—monatomic, diatomic, tetratomic, and polyatomic molecules—corresponding anomalies in the condensing powers of the respective atoms of these bodies—physical aids to the determination of atom-weights—specific-heat method—isomorphic method—cases in which they are of special value—table of the elements, with their atomic and molecular symbols, weights, volumes (so far as known) and coefficients of quantivalence—one exception to the universality of the symbols—appeal, on this subject, to scientific France.

Thus far, in our study of the relations of the elements, as to their combining proportions, volumetric and ponderal, and as to their chemical powers, molecule-forming and atom-fixing, we have confined our attention to a few typical bodies and their immediate analogues. Hydrogen has stood apart as our unitary standard; while chlorine, oxygen, nitrogen, and carbon, each grouped with two congeners—thirteen bodies in all—have supplied us with the whole of the facts we have employed, to form the basis of our demonstration, and to establish that symbolic language which, expanding as we proceeded, has served both to facilitate our investigations, and tersely to embody our results.

We have now to step beyond this limited sphere of inquiry,

and to determine how far our methods of research, our principles of classification, and our symbolic system of record, derived from the study of thirteen elements only, are applicable to the investigation and co-ordination of the sixty-one elements which constitute our planet.

For such of the typical elements and their congeners as are permanently gaseous, or readily vaporizable, we have had no difficulty in directly fixing the relative gas-volume weights or vapour-densities, by weighing one litre of each gas, or vapour, in comparison with the corresponding measure of hydrogen taken as unity. These relative gas-volume-weights or specific gravities of the volatile elements, we have found generally to represent their minimum molecule-forming weights; also designated their atom-weights, or combining numbers.

But the exceptional cases of phosphorus and arsenic, whose gas-volume-weights or specific gravities we found to be double their molecule-forming or atom weights, obliged us to regard the specific gravity of gaseous bodies as but an uncertain criterion of their combining numbers; though, on the other hand, the extreme simplicity of the ratio (2 to 1) subsisting between the specific gravity and atom-weight of phosphorus and arsenic respectively, afforded us a good prospect, if not of certainly making out, by the specific gravity method, the true combining atom-weights of volatile elements, at all events of arriving at figures standing in very simple ratios thereto, and therefore likely, if taken in conjunction with other indications, to assist in guiding us to the desired result.

The method upon which, in the case of phosphorus and arsenic, we were compelled to fall back, for the determination of their true atom-weights, consisted in ascertaining, by means of the balance, the weight of each of those bodies, respectively, contained in the normal product-volume (2 litres) of the compound gas formed by it with hydrogen.

In the case of carbon we resorted to the same method for a different reason, viz., the impossibility of volatilizing carbon, so as to ascertain directly its relative gas-volume weight. We therefore determined, by means of the balance, the proportion by weight of carbon contained in two litres of that one of its gaseous compounds with hydrogen in which it is least abundant; and this weight we took, on the ground of analogy, as the true molecule-forming minim, or combining weight, otherwise called atom-weight, of carbon.

So far as the other non-volatile elements are obtainable in gaseous combination with hydrogen, the method which thus served us, in the case of carbon, is evidently applicable, in their cases also, to determine their respective molecule-forming minims, or atom-weights.

Unfortunately, however, a large majority of the elementary bodies, while resembling carbon as to their non-volatility under any treatment in our power to employ, differ from carbon in the circumstance that they have hitherto resisted our endeavours to bring them into combination with hydrogen.

The problem now before us is, therefore, how to determine the molecule-forming minims, or atom-weights, of those numerous bodies which neither yield vapours to be directly weighed, nor form gaseous or volatile compounds with our unitary standard, hydrogen?

It is a well-known mathematical axiom, that "things which are equal to the same thing are equal to one another:" and this axiom is as true with respect to chemical as it is with respect to mathematical quantities.

A little consideration will satisfy us that this self-evident principle, coupled with the knowledge which we now possess, of molecular and atomic values, greatly enlarges the otherwise restricted range of facts, available as sources of information concerning the atom-weights of the elements still remaining to be investigated.

Thus, knowing as we do the molecule-forming minim-weight, and the atom-fixing or quantivalent power, of chlorine, relatively to hydrogen taken as unity; and knowing also that the product-volume of volatile chlorine-compounds corresponds precisely with that of volatile hydrogen-compounds, such volume being, for both classes of compounds, dilitral; it is clear that, for elements which do not enter into combination with hydrogen, but which do form volatile or volatilizable compounds with chlorine, we may fall back on these chlorine-compounds, and, from the ponderal analysis thereof, deduce the molecule-forming minims or atom-weights sought, as readily and certainly as if the missing hydrogen-compounds themselves were at our disposal.

The axiom which we have quoted as placing chlorine on a par with hydrogen for the purpose in view, is, of necessity, equally applicable (*mutatis mutandis*) to show oxygen and nitrogen available to the same end.

Of course, in so using oxygen and nitrogen, we must take duly into account their respective molecule-forming minim-weights and atomic quantivalence, as to both of which values, these gases (we know) differ, both from each other and from chlorine. But, these being allowed for, the volatile oxygen and nitrogen compounds of the non-volatile elements are as available as their corresponding chlorine-compounds, nay, even as their hydrogen-compounds themselves (when such exist), for the determination of their molecule-forming minims or atom-weights.

Among the experiments which we have already made there happens fortunately to be a series capable of furnishing appropriate and satisfactory illustrations of our present theme.

We employed, as you well remember, the alkali-metals, sodium (Na) and potassium (K), to withdraw chlorine from hydrochloric acid (HCl^{i}), oxygen from water (H_2O^{ii}), and nitrogen from ammonia (H_3N^{iii}).

The compounds of Na and K which we thus formed respectively with chlorine, with oxygen, and with nitrogen, we were led, by analysis, to represent in formulæ, which displayed most clearly their exact structural analogies with the three hydrogen compounds—hydrochloric acid, water, and ammonia—from which they respectively originated.

In the following table we have placed in the first column the names, symbols, coefficients, and atom-weights of chlorine, oxygen, and nitrogen; in the second column, the formulæ of their respective combinations with hydrogen; and in the third and fourth columns respectively, the formulæ of their analogous combinations with sodium and potassium. Thus, reading the table horizontally, we have, in the first row three *chlorides*, in the second row three *oxides*, and in the third row three *nitrides*, brought into direct comparison.

COMPOUNDS OF CHLORINE, OXYGEN, AND NITROGEN, WITH HYDROGEN, SODIUM, AND POTASSIUM.

Names, symbols, coefficients of quantivalence, and combining weights of the three elements existing in all the Compounds formed.	Names, symbols, coefficients of quantivalence, and combining weights of the elements taking part, each in three of the Compounds formed; with formulæ of the products.		
	Hydrogen H = 1.	Sodium Na = 23.	Potassium K = 39.
Chlorine = Cl^{I}, 35·5	HCl^{I}	$NaCl^{I}$	KCl^{I}
Oxygen = O^{II} 16.	H_2O^{II}	Na_2O^{II}	K_2O^{II}
Nitrogen = N^{III} 14.	H_3N^{III}	Na_3N^{III}	K_3N^{III}

Now, H representing unity as a matter of course, why have we in this table made Na = 23 and K = 39 respectively? Because these are the respective weights of those two bodies which, in each of the compounds figured in the table, exactly replace, and do the work of, 1 part by weight of hydrogen.

Carrying the eye along the chlorine line, we see that one part by weight of hydrogen in HCl^{I} is replaced, and its atom-fixing duty fulfilled, by 23 parts by weight of sodium in $NaCl^{I}$, and by 39 parts by weight of potassium in KCl^{I}; or, to express the same thing more succinctly, we see Cl^{I} = 35·5 satisfied, equally and indifferently, by H = 1, by Na = 23, and by K = 39.

Again, carrying the eye along the oxygen line, we see $H_2 = 2$ replaced, and its atom-fixing function performed by Na_2, or $23 \times 2 = 46$; and by K_2, or $39 \times 2 = 78$.

Lastly, glancing similarly along the nitrogen line of the table, we see $H_3 = 3$ replaced by Na_3, or $23 \times 3 = 69$; and by K_3, or $39 \times 3 = 117$.

Short of a demonstration by actual experiment, it is difficult to conceive stronger grounds than those afforded by these comparisons, for regarding Na and K as univalent bodies, = Na^{i} and K^{i}, and assigning to them respectively 23 and 39, as their atomic weights.

Happily, as both Na^{i} and K^{i} are volatile metals, and as their chlorides and oxides are also volatile (at furnace temperatures), we may hope to obtain, in due time, direct experimental proof that their respective atom-weights are 23 and 39, and that their atoms are univalent.

Such proof would be evidently obtained, should we succeed in determining the weights of the product-volumes of $Na^{i}Cl^{i}$ and $K^{i}Cl^{i}$, and should we find these weights equal to $23 + 35{\cdot}5 = 58{\cdot}5$ for $Na^{i}Cl^{i}$, and to $39 + 35{\cdot}5 = 74{\cdot}5$ for $K^{i}Cl^{i}$. These proofs would, of course, be powerfully corroborated should we further succeed in weighing the dilitral volumes of $Na^{i}_{2}O^{ii}$ and $K^{i}_{2}O^{ii}$, and should these weights prove, conformably with our anticipations, to be $23 \times 2 + 16 = 62$ for the former oxide, and $39 \times 2 + 16 = 94$ for the latter.

It is hardly necessary to add, that assurance would be made doubly sure, should we at any future time succeed in obtaining volatile hydrogen-compounds of sodium and potassium, and should these prove, on analysis, to have, respectively, the composition and weight, $HNa^{i} = 1 + 23 = 24$, and $HK^{i} = 1 + 39 = 40$, in one dilitral product-volume (comp. p. 199).

In the present eminently transitional state of chemistry we do not possess any such absolutely irrefragable proofs of the atom-weights of Na^{i} and K^{i}; and, pending the researches which, in a hundred laboratories, are constantly in progress for the more strict solution of these and many cognate problems, we provisionally adopt, not for Na^{i} and K^{i} only, but for numerous other elements, atom-weights resting on strong grounds of probability, but still awaiting their final and unconditional verification.

For bodies which appear likely to resist, in the future as in the past, all attempts to obtain and weigh the vapour, either of themselves or of any of their compounds, we must of course expect to rely permanently, for the determination of their respective atom-weights, upon the ponderal analysis of their fixed combinations with one or more of the typical elements (as, for instance, with chlorine and oxygen); the minim-weight which combines with 35·5 of univalent chlorine, and half the minim-weight which combines with 16 of divalent oxygen, being adopted, in each such case, as the atom-weight sought. Atom-weights thus determined exclusively by aid of the balance must, of course, remain permanently liable to the same uncertainty as provisionally attends like purely ponderal determinations of the atom-weights of bodies which, like Na' and K', we hope hereafter to submit to volumetric verification.

A few examples, chosen from among the elements to which our attention has not yet been turned, will assist us in firmly grasping the method of fixing, both by ponderal and volumetric determinations, the respective molecule-forming minims, or atom-weights of those elements, and their atomic quantivalence.

For this purpose we will select three familiar metals: Mercury, with which our pneumatic trough is filled, and which we see in every looking-glass and every barometer; Bismuth, an ingredient of common pewter and of the well-known fusible alloy; Tin, with which our kitchen implements are coated.

Neither of these metals has yet been obtained in combination with hydrogen; but they all of them combine readily with chlorine; and, fortunately, their compounds with chlorine are volatile, so that we are enabled to determine their vapour-densities.

These have been accurately ascertained; as also have the proportions by weight in which their respective constituents are combined therein. These facts are set forth in the following table:—

	Gas-volume-weight.	Proportions of the Constituents therein	
		Metal.	Chlorine.
Chloride of mercury	135·50	100	35·5
„ bismuth	157·25	104	53·25
„ tin	130·	59	71·

From this table it is of course easy to compute the weight and composition of the normal product-volume of each of these compounds. For, this volume being, as we know, dilitral, while the unit-volume referred to in the above table is monolitral, we have only to multiply the above figures by 2 to obtain the corresponding dilitral, or, as we now say, the molecular expressions.

These calculations have been made, and they give us the following results.

The dilitral or product-volume of chloride of mercury is found to contain 2 unit-volumes, representing (as we now know) 2 atoms, of chlorine; or, by weight 35·5 × 2 = 71 parts of chlorine; from which fact we at once learn that the atom of mercury is bivalent. With these 71 parts of chlorine are associated, in the product-volume of mercurial chloride, 200 parts by weight of mercury; which fact acquaints us with the atom-weight of mercury, whose Latin name is *hydrargyrum*. To this metal we accordingly assign the symbol Hg^{II} = 200.

Again, the dilitral or product-volume-weight of chloride of bismuth contains 3 unit-volumes, representing 3 atoms, of chlorine; or, by weight, 35·5 × 3 = 106·5 of chlorine; whence we learn that bismuth is trivalent. With these 106·5 parts of chlorine are associated, in the product-volume of the bismuth-chloride, 208 parts by weight of bismuth; bespeaking, for the bismuth-atom, the symbol and weight Bi^{III} = 208.

Lastly, in the dilitral or product-volume of tin chloride, analysis reveals 4 unit-volumes = 4 atoms of chlorine; or, by weight, 35·5 × 4 = 142 of chlorine; proving tin to be quadrivalent, like carbon. With these 142 parts of chlorine we find associated in the product-volume of the tin-chloride, 118 parts of tin, in Latin *stannum*; whose atom we accordingly symbolize as Sn^{IV} = 118.

This method is of universal applicability; indeed, of all the methods available for use it is the one most free from liability to error. It implies, however, the assumption that the product-volume in each case contains *one* atom only, not two or more atoms, of the metal under examination. Further data are, therefore, necessary in these cases, in order that this doubt may be set at rest. And such data are fortunately supplied by methods to which we shall have occasion to refer in the sequel. (Comp. pp. 199—200.)

We have already seen, in the cases of phosphorus and arsenic, that the relative volume-weights of the volatile elements, though they usually coincide with the atom-weights, are liable to deviate therefrom exceptionally; so that they cannot be relied on as certain indices of the molecule-forming minims.

Further proof of this fact is afforded by one of the three metals just examined, viz., mercury—the only one of the three, as it happens, whose vapour-density has been determined. The vapour-density of mercury does not coincide with its atom-weight or molecule-forming minim. Like phosphorus and arsenic, mercury is in this respect an exception to the usual rule. Curiously enough, however, the deviation from type in the case of mercury (which the metal Cadmium in this respect resembles) is in precisely the opposite direction from that in which phosphorus and arsenic diverge therefrom.

While the atom-weights of phosphorus (P^{III}) and arsenic (As^{III}) express respectively only *half* the vapour-densities of these elements, the atom-weights of mercury (Hg^{II}) and cadmium (Cd^{II}) represent, on the contrary, *double* the vapour-densities of those bodies respectively. As, therefore, for the correct symbolization of the P^{III} and As^{III} atoms, we had to *halve* our usual monolitral squares, so, to represent faithfully the atoms of Hg^{II} and Cd^{II}, we have now to employ a *double* square—our ordinary dilitral symbol.

For the clearer comprehension of these facts they are set forth collectively in the following diagrammatic and descriptive table:—

SYMBOLIZATION of CERTAIN ELEMENTS, STANDARD & EXCEPTIONAL.

1. Standard.

Hydrogen = unity.

Characters :—

(*a*.) Atom-weight *equal* to specific gravity.

(*b*.) Molecular structure *diatomic*.

Symbols :—

Atomic (*monolitral*).	*Molecular* (*diatomic and dilitral*).
H	HH

2. Exceptional.

α. Phosphorus and Arsenic.

Characters :—

(*a*.) Atom-weight equal to *half* specific gravity.

(*b*.) Molecular structure *tetratomic*.

Symbols :—

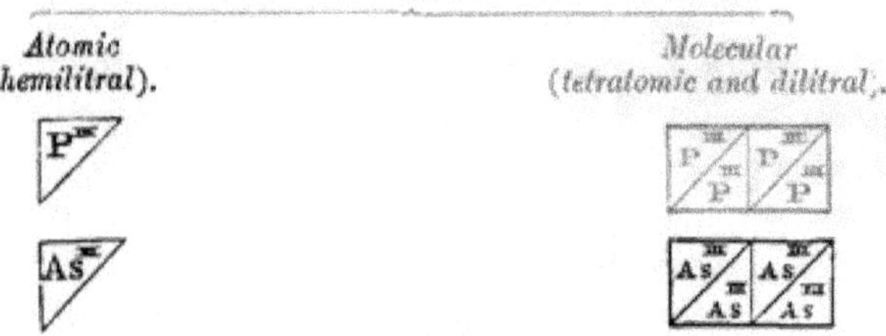

β. Mercury and Cadmium.

Characters :—

(*a*.) Atom-weight equal to *twice* specific gravity.

(*b*.) Molecular structure *monatomic*.

Symbols :—

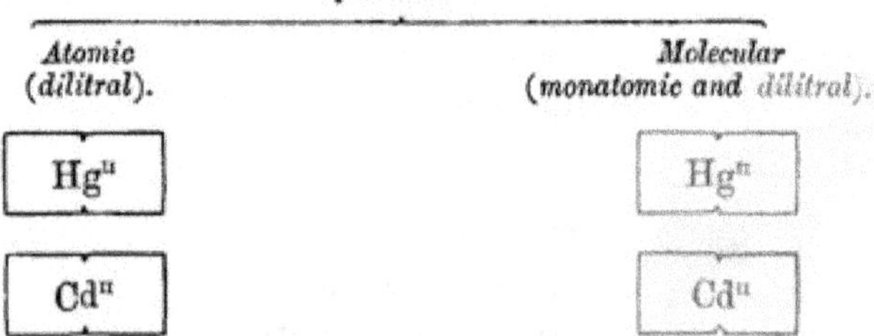

The most remarkable feature in this table is the coincidence it brings out between the *atom* and the *molecule* of the respective metals, mercury and cadmium; whose molecular structure is thus represented as *monatomic*. On this point, however, we must be on our guard against attributing to the results of a particular system of symbolization a deeper significance than they really possess. It is only in our hypothetical conception, not as a result of experimental determination, that the molecule of hydrogen is regarded as diatomic. All that we really know of the structure of hydrogen is, that its free molecule, HH, represents double the value of its combining particle, H, in point of weight, magnitude, and numerical complexity of structure. We assume H to be an ultimate atom merely for simplicity's sake. The smallest quantity in which hydrogen combines may, for aught we know to the contrary, be a cluster numbering ten, or a hundred, or a thousand, or a million, of really ultimate atoms; and consequently, the molecular symbol HH merely implies that, whatever number of ultimate atoms there may be in H, from 1 upwards, the free hydrogen molecule contains twice that number. Or, putting it in algebraic form, if n represent the numerical constitution of the molecule HH, then the similar constitution of the (so called) atom, H, is represented by $\frac{n}{2}$. Similarly, the molecular structure of mercury, Hg, and of cadmium, Cd, relatively to that of $HH = n$, is for mercury $Hg = \frac{n}{2}$ and for cadmium $Cd = \frac{n}{2}$; so that it is but the assumption that the hydrogen molecule is *di*atomic, which obliges us to describe the mercury and cadmium molecules as *mon*atomic. These molecules may with equal probability be assumed to be of million-fold complexity, if only care be taken to raise in a proportionate degree our conception of the respective complexities of the so-called diatomic and tetratomic molecules.

We need hardly point out that, in the exceptional cases of mercury and cadmium, the volume-condensing power of the atoms deviates from its ordinary relation to the quantivalent power; and that this deviation is in the sense opposite to that which we have already pointed out in the cases of phosphorus and arsenic. As the atoms of trivalent phosphorus and arsenic are of only *half* the normal volume, the product-volumes of their respective compounds are (so to speak) less crowded, by exactly half a volume, than the product-volumes of the corresponding compounds of their prototype, trivalent nitrogen. By parity of reasoning, as the atoms of bivalent mercury and cadmium are of *double* the normal volume, the product-volumes of their respective compounds are packed (so to speak) with one volume more than the product-volumes of the corresponding compounds of their prototype, bivalent oxygen.

These anomalies are, no doubt, troublesome, as impairing the unity of our symbolic language, and burdening the memory with several exceptional details. With the foregoing explanations, however, they cannot lead us to form erroneous conceptions—and this is the principal evil to be avoided.

As for their philosophical interpretation, this belongs to the future. They may turn out to be typical facts, round which many others of like kind may come hereafter to be grouped; and they may prove to be allied with special properties, or dependent on particular conditions, as yet unsuspected.

We shall heartily welcome, as we must patiently await, such explanations hereon as time may have in store; merely making it our care, meanwhile, faithfully to record the facts as they stand.

From the foregoing remarks it appears that, even for elements which are volatile, and of which the gas volume-weights (or specific gravities) can be immediately determined, this direct method of ascertaining their respective molecule-forming minims, or atom-weights, is subject to error, on account of their liability to such anomalies as we have met with in the cases just cited.

It is by first obtaining the gaseous or volatile compound formed by an element, either with hydrogen itself, or, failing this, with some element bearing thereto known weight and volume ratios, and by then ascertaining its vapour-density and composition, so as to learn how much of the element under investigation exists in the dilitral volume of such compound, that the atom-weights of the elements, whether volatile or fixed, can be most safely and certainly determined.

To this, the purely chemical method, certain physical modes of research afford, in doubtful cases, very acceptable aid.

Much light, for example, is thrown on the atom-weights of the elements by reference to their Specific Heats; *i. e.*, to the number of units of heat which proportionate weights thereof absorb in acquiring a given sensible temperature (the unit of heat being so much as suffices to raise 1 litre of water 1 degree centigrade). Now it has been found that, though equal weights of the elements absorb very different quantities of heat in attaining to equal sensible temperatures, and therefore possess very unequal specific heats, as thus measured; yet if, of each element compared, a quantity equal to its atom-weight be taken, the absorption of heat, in attaining the same temperature, becomes precisely equal for all; or, to express the same thing in other words, the atom-weights of the elements, multiplied into their specific heats, give a constant product. This law is subject, unfortunately, to exceptions; otherwise it would afford a very simple and universally applicable mode of ascertaining the atom-weights of elementary bodies. Even as it is, most valuable preliminary hints for guidance, and welcome corroborations of chemically-obtained results, are frequently afforded by these physical determinations; and the case of sodium, dwelt on above, may be cited as one of those, in which the specific-heat investigation confirms the indication of the atom-weight, derived from purely chemical research (comp. p. 192).

It is remarkable that the specific-heat method often affords the most distinct indications as to the atom-weights of elements, precisely in the cases in which such assistance is most needed; as, for instance, in the cases of the non-volatile elements, referred to above; and also in cases which, notwithstanding that volatile compounds of the elements whose atom-weights are sought, exist and have been weighed, are rendered doubtful by the uncertainty whether one, two, or more of the atoms in question are contained in the respective product-volumes. Uncertainty of this kind prevailed, as we have pointed out (comp. p. 194), in the cases of mercury, bismuth, and tin, until it was set at rest by reference to their specific heats; which confirmed the view that the product-volumes of their chlorides contain but one atom of the respective metals. There is evidently still greater doubt to set at rest in cases where the only volatile bodies available for the determination of the required atom-weights, by the vapour-density method, form *series*, whereof all the members contain the same elements united in various multiple proportions. In these cases also, two or more possible atom-values offer themselves for selection, often with very evenly-poised claims to preference; and it is then that the determination of the specific heat comes in, with welcome weight, to turn decisively the otherwise doubtful balance.

The phenomena of Crystallization afford similar assistance, in a still larger number of doubtful cases, by revealing between bodies certain chemical analogies which, in their turn, elucidate questions as to their atom-weights. We shall hereafter learn that bodies of similar atomic constitution are very commonly *isomorphic*, *i. e.*, that they affect similar, or closely-related, crystalline forms. Hence the comparison of two crystalline bodies, one of doubtful, the other of well-ascertained atomic construction, will often afford conclusive information as to an atom-value else uncertain.

It forms no part of our present plan to enter into the details of these physical modes of research, nor to exemplify their application to particular cases. Indeed, we have not yet studied a sufficient number of the elements and their compounds to prepare us for pursuing this line of research; which we shall have better opportunities of taking up hereafter, when further advanced in our studies.

For the present we may close this branch of our inquiry by subjoining a general table of the 61 elements, with their names alphabetically arranged in the first column; their symbols and coefficients of quantivalence, with their atom-weights and volumes in the second; and their molecular symbols, weights, and volumes, in the third.

The molecular values, and all the volumetric values, atomic as well as molecular, are, it will be observed, given in this table for the volatile elements only; and not for all even of these. The blanks await filling up with the results of future experiment; let us hope at no distant date. It is further to be observed, with reference to this table, that a certain number even of the atom-weights therein assigned, especially to the rarer and less perfectly studied elements, are still matters in question among chemists. These doubts are most prevalent in the cases (already alluded to) of elements which combine in multiple proportions to form series of compounds, from which this or that may be chosen, at pleasure, to serve as the basis of the atom-weight determination. In the present transitional state of chemistry, indeed, many of the atom-weights must be taken as provisional, and subject to verification. In studying the individual elements, we shall take occasion to set forth the experimental facts, and the considerations founded thereon, which have led to the adoption of each of the figures given in the table; and, perhaps, as we proceed, it may fall within our power to correct some of the doubtful values, and to fill up some of the blanks. These remarks apply not only to the molecule-forming weights, but also to the coefficients of quantivalence by which their atom-fixing powers are expressed.

The use of these last-mentioned exponents may be once more mentioned in connection with this table.

In order to ascertain what weight of each element suffices to fix or replace one standard univalent atom (H = 1), or a proportionate part of any atom of higher quantivalence, we have only to divide the atom-weight of any body by its quantivalential coefficient.

The atom-weight of sulphur, for example, is shown, in column 3, to be 32; while its coefficient of quantivalence is given in the column of literal symbols as S^{II}. Therefore the quantity of sulphur requisite to replace 1 of hydrogen is $\frac{32}{2} = 16$

With these explanations as to its employment, and subject to these reserves as to its accuracy, this table will be found, I trust, to convey, in a simple and compendious form, a considerable mass of useful information.

NAMES	ATOMS.			MOLECULES.		
of the Elements, fixed and volatile, arranged in alphabetical order.	Ponderal Symbolization of *all* the Elements.		Pondero-volumetric Symbolization of the *Volatile* Elements only.			
	SYMBOL, (literal) and coefficient of quantivalence.	WEIGHT, atomic, or combining; representing the molecule-forming minim.	VOLUME, Diagrammatic representation of; showing the relative size of the combining atom.	SYMBOL (literal) representing the atomic structure of the free molecule.	WEIGHT of the free molecule; or free gaseous minim-weight.	VOLUME, Diagrammatic representation of; showing the atomic structure of the free molecule.
(*Standard = unity*) HYDROGEN	H^{I}	1	H	HH	2	HH
Aluminium	Al^{III}	27·5				
Antimony	Sb^{III}	122				
Arsenic	As^{III}	75	As	AsAsAsAs	300	As As As As
Barium	Ba^{II}	137				
Beryllium	Be^{III}	14				

Bi^{III}	208				
Br^{I}	80	Br	BrBr	160	BrBr
Bo^{III}	11				
Cd^{II}	112	Cd	Cd	112	Cd
Cs^{I}	133				
Ca^{II}	40				
C^{IV}	12				
Co^{II}	92				
Cl^{I}	35·5	Cl	ClCl	71	ClCl
Cr^{III}	52·5				
Co^{II}	58·8				
Cu^{II}	63·5				
Di^{II}	96				
F^{I}	19				
Au^{III}	196·7				

NAMES of the Elements, fixed and volatile, arranged in alphabetical order,	ATOMS.			MOLECULES.		
	Ponderal Symbolization of *all* the Elements.		Pondero-volumetric Symbolization of the *Volatile* Elements only.			
	SYMBOL, (literal) and coefficient of quantivalence.	WEIGHT, atomic, or combining; representing the molecule-forming minim.	VOLUME. Diagrammatic representation of; showing the relative size of the combining atom.	SYMBOL (literal) representing the atomic structure of the free molecule.	WEIGHT of the free molecule; or free gaseous minim-weight.	**VOLUME, Diagrammatic representation** of; showing the atomic structure of the free molecule.
Hydrogen	H^{I}	1	H	HH	2	HH
Iridium	Ir^{IV}	198				
Iodine	I^{I}	127	I	II	254	II
Iron	Fe^{II}	56				
Lanthanium	La^{II}	92				
Lead	Pb^{II}	207				

Lithium	Li^{I}	7				
Magnesium	Mg^{II}	24				
Manganese	Mn^{II}	55				
Mercury	Hg^{II}	200	Hg	Hg	200	Hg
Molybdenum	Mo^{VI}	92				
Nickel	Ni^{II}	58·8				
Niobium	Nb^{IV}	97·6				
Nitrogen	N^{III}	14	N	NN	28	NN
Osmium	Os^{IV}	199				
Oxygen	O^{II}	16	O	OO	32	OO
Palladium	Pd^{II}	106·5				
Phosphorus	P^{III}	31	P	PPPP	124	P P P P
Platinum	Pt^{IV}	197·4				
Potassium	K^{I}	39				

THE ELEMENTS: **their Nomenclature and** Symbolization—*continued.*

NAMES of the Elements, fixed and volatile, arranged in alphabetical order.	ATOMS.			MOLECULES.		
	Ponderal Symbolization of *all* the Elements.		Pondero-volumetric Symbolization of the *Volatile* Elements only.			
	SYMBOL, (literal) and coefficient of quantivalence.	WEIGHT, atomic, or combining; representing the molecule-forming minim.	VOLUME, Diagrammatic representation of; showing the relative size of the combining atom.	SYMBOL (literal) representing the atomic structure of the free molecule.	WEIGHT of the free molecule; or free gaseous minim-weight.	VOLUME, Diagrammatic representation of; showing the atomic structure of the free molecule.
Rhodium	Rh^{II}	104				
Rubidium	Rb^{I}	85·5				
Ruthenium	Ru^{IV}	104				
Selenium	Se^{II}	79	Se	SeSe	158	SeSe
Silver	Ag^{I}	108				
Silicon	Si^{IV}	28·5				
Sodium	Na^{I}	23				

Strontium	Sr^{II}	87·5				
Sulphur	S^{II}	32	S	SS	64	SS
Tantalum	Ta^{IV}	137·5				
Tellurium	Te^{II}	128				
Thallium	Tl^{I}	204				
Thorium	Th^{IV}	231·5				
Tin	Sn^{IV}	118				
Titanium	Ti^{IV}	50				
Tungsten	W^{VI}	184				
Uranium	U^{II}	120				
Vanadium	V^{VI}	137				
Yttrium	Y^{II}	68				
Zinc	Zn^{II}	65				
Zirconium	Zr^{IV}	90				

One conspicuous merit of the elementary symbolization epitomized in this table consists in its Universality; I mean in its general recognition and employment by the chemists of all nations throughout the civilized world.

From this broad statement, however, we have to except one symbol, and one nation. That symbol is the expression N = nitrogen; and that nation is France; for, France alone, among all the nations of the earth, continues to designate nitrogen by the symbol and name Az = azote. It is time that this single deviation from uniformity in the elementary chemical language of the world should disappear; and I confidently appeal to our scientific brethren beyond the Channel to efface this one remaining discrepancy by adopting the symbol N.

It is not necessary, for this purpose, to discontinue the use of the name *azote* also. The Italians employ the term *azoto*, but they are beginning to couple with it the symbol N. In the cases of the alkali metals, the French themselves concur with other nations in using the symbols K and Na, though these are not the initial letters of the names (*potasse* and *soude*), which they assign to the alkalies. Evidently, therefore, there would be no logical inconsistency involved in this slight concession, on the part of France, to the general usage of the scientific comity of nations. On the contrary, France owes this rectification quite as much to her own genius for philosophical order and unity, as to that general European sentiment whereof, in this appeal, I am but the humble spokesman. Moreover, to a similar, and much more sweeping appeal, on the part of France, scientific Europe has responded with alacrity, by renouncing, in favour of the noble French metrical system, a thousand incongruities of local weight and measure. And again, in the great work of building up, on a unitary basis, the magnificent edifice of modern chemistry, France has taken, in the past as in the present, her fully proportionate share. These are all valid and logical arguments in support of our present plea. When, therefore, we solicit at her hands the abandonment of her one discordant symbol, we do but ask her to carry out the spirit of her own unitary conceptions; and to the successors of Lavoisier—to the contemporaries of Gerhardt—our appeal, we are persuaded, will not be made in vain.

LECTURE XII.

Compounds of a higher order, ternary, quaternary, &c.—laws of proportionality and quantivalential relations exemplified in the generation of such compounds—their frequently high vapour-densities—their tendency, in many cases, to *dissociation*—examples of ternary compounds—hydrochlorate of ammonia—its production by the union of the molecules of two gaseous binary compounds—its neutral, salt-like characters—dissociation of its vapour—ternary compounds produced during the progressive dehydrogenation of water and ammonia by the alkali-metals—unisodic water, or hydrosodic oxide—unisodic and bisodic ammonias, or hydro-unisodic and hydro-bisodic nitrides—progressive expulsion and replacement of hydrogen-atoms by sodium-atoms during the formation of these compounds—analogous substitution-compound in the marsh-gas group—general conception of substitution-compounds—their production further illustrated in the progressive transformation of water, ammonia, and marsh-gas, by inception of chlorine and expulsion of hydrogen—retention of the structural type of the parent compound by its substitutional derivatives—conversion of binary into ternary compounds by atomic inception, unattended by substitutional displacement—exemplification of this mode of their genesis in the hydrochloroxygen series—analogous ternary compounds of the hydrochloric group—of the water group—hydrosulphoxygen series—of the ammonia group—hydrophosphoxygen series—of the marsh-gas group—methylic alcohol—its transitional importance—close of the lecture—summary recapitulation and conclusion of the course.

The principles of symbolization, literal and diagrammatic, illustrated in the table of the elements to which our attention was directed at the close of the last lecture, apply equally to the concise representation of the structure of compound bodies.

Thus, as, in the table, we denote the elementary hydrogen molecule by the letters HH, and depict diagrammatically its diatomic structure by the dilitral figure—

HH

so, likewise, we represent the similarly-formed diatomic compound molecule of hydrochloric acid by the letters HCl^{i}, and the dilitral figure—

HCl^{i}

Again, as in point of structure and dimensions the polyatomic binary molecules of water-gas, ammonia-gas, and marsh-gas, are all precisely analogous to the elementary diatomic standard molecule, HH, we represent these also, respectively, in the literal expressions H_2O^{ii}, H_3N^{iii}, and H_4C^{iv}, and by the dilitral double-square symbols—

When we wish to represent elementary or compound molecules, whose dimensions have not been ascertained by experiment, but are provisionally assumed on analogical grounds, we resort to the dotted lines previously employed by us in like cases; representing, for example, the hypothetical gaseous molecules of the alkali-metals, and their compounds with chlorine and oxygen, as follows:—

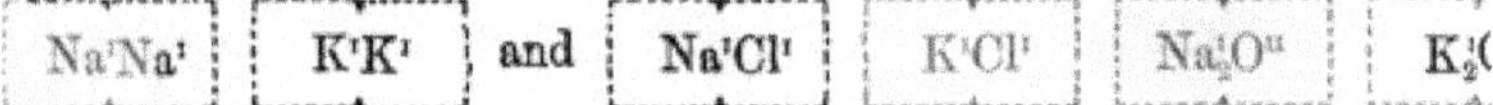

Though we have hitherto had occasion to employ these methods of symbolization only for the representation of the elements and their binary compounds, they are equally available (*mutatis mutandis*) to denote the composition and structure of compounds of a higher order; that is to say, of compounds in whose formation not two only, but three, four, five, and sometimes even six elements take part, and which are accordingly distinguished as ternary, quaternary, quinternary, and sexternary compounds.

Of these compounds of a higher order we have as yet refrained from speaking; though, in the course of our experiments, —even of the very first we made—several such bodies passed through our hands. The fact is that, in those earlier stages of our inquiry, reference to these bodies would have been premature and purposeless; a mere incumbrance of the memory with details not contributing to the elucidation of the subjects then in hand. The time, however, is now come when we may, with advantage, pick up these dropped links and add them to our chain. The study of their more complex structure will prepare us for the investigation, upon which we must hereafter enter, of an almost infinite succession of complex bodies, so numerous and so ever multiplying, that he who desires to be a master in chemistry must be content to remain, his whole life through, a hard-working student of their legions.

We may premise briefly, with respect to these compounds of a higher order, that they are generated by the operation of the same forces, exercised in accordance with the same laws, as determine the production of the binary compounds. The activity of those forces is as often attended, in the one as in the other class of cases, by perturbations, manifested in the development of light and heat. The laws of proportionality are, moreover, as exactly fulfilled in the formation of the most complex of these higher compounds as in that of the simplest molecule; each atom that takes part in a combination invariably engaging therein either in the single or multiple ratio of its weight, and carrying with it unchanged, throughout the most complicated reactions, its peculiar atom-fixing or quantivalential powers.

Hence it results that, in the formation of many of these compounds of a higher order, the dilitral product-volume, into which we have as yet seen condensed only some half dozen volumes at most, becomes packed, so to speak, with scores of volumes. Hence we may readily conceive that, in the study of the more complex, as of the simpler bodies, the product-volume is still (whenever they are volatile) the starting-point of the investigation; that is to say, we begin by determining the ponderal composition and vapour-density of the body under examination, and thence deduce the weight and constitution of its dilitral volume, which corresponds, as we know, with the weight and structure of its molecule.

This method is, however, subject in all cases to a difficulty on which we have not yet touched, and which it is now opportune to point out and illustrate.

It frequently happens that our attempts to determine the vapour-densities of volatile compounds, whether binary or of a higher order, are baffled by their splitting up, at the temperature of vaporization, into their constitutent gases or vapours; which, of course, how much soever they may have been condensed during combination, resume, in the act of separating, their proper volumes.

We need not look far, either for examples of compounds of a higher order, or for illustrations of this splitting up, or, as it is termed, *Dissociation*, of volatile bodies, under the influence of elevated temperatures.

Among the experiments we have made here together, you doubtless remember one in which, when acting on ammonia with chlorine gas (comp. p. 56), we incidentally produced a white crystalline deposit on the interior of the tube we employed. That deposit we laid aside at the time, as having no bearing on the question then under consideration, viz., the volume-ratio of the two gaseous constituents of ammonia. But now this deposit acquires particular interest for us, firstly, as being a ternary compound, and, secondly, as exemplifying the phenomenon of *dissociation*.

We will, therefore, fix our attention on this body; and, for this purpose, we will begin by producing a small quantity of it; a result which we may readily accomplish by simply bringing its constituent gases into contact with each other.

With the experience in manipulation we now possess, we can be at no loss to perform this easy operation. Here is our familiar pneumatic apparatus, our mercurial trough, with its inverted cylinders, for the reception and admixture of the gases under examination. A pair of these cylinders, of equal size, have been already filled, by the means we have so often employed, the one with dry ammonia-gas, the other with dry hydrochloric-acid gas. Beside these receptacles, a third, of larger size, stands inverted over the trough; but this one is filled to the top with mercury, in readiness for the reception of gas. Into this cylinder we first pass, by upward decantation, the hydrochloric acid gas (marking the space it fills by a caoutchouc ring), and then, bubble by bubble, we add the ammonia.

Fig. 65.

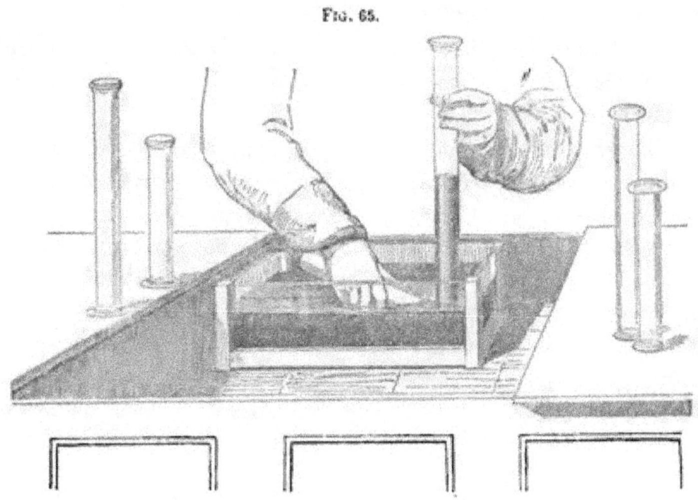

As each bubble of this gas comes into contact with the hydrochloric acid, a dense white cloud is seen to form; and this condenses into crystalline flakes, which collect on the interior of the vessel. During this process the original volume of the hydrochloric acid gas gradually diminishes. Indeed, the more ammonia we pour up through the mercury, the higher does this metal rise in the cylinder; and with the last bubble of ammonia sent up, the last trace of the gaseous mixture disappears, and the vessel now only contains, beside mercury, the film of white deposit into which the two gases have condensed.

We are, therefore, justified in regarding this deposit as a compound of the two gases we have mixed, viz., hydrochloric acid and ammonia, united in equal volumes.

As a corroborative demonstration of this fact, we have here the means of mixing the two gases as before, but with excess of either gas at pleasure. Into this cylinder we introduce hydrochloric acid gas in excess of the ammonia; into this other one, ammonia in excess of the hydrochloric acid: in both cases, you observe, combination takes place as before; but in both cases there is a surplus of uncombined gas left. By means of these test papers we further learn that the surplus consists, in one case, of ammonia; in the other, of hydrochloric acid—a clear proof that combination has only taken place between equal volumes of the two gases.

The white, solid product we find, upon trial, to be readily soluble in water; and, on comparing its solution with solutions of its respective constituents, we find that the characteristic properties of these latter have entirely disappeared in the product of their union. The suffocating exhalation of the hydrochloric acid, the pungent odour of the ammonia, are no longer perceptible in the solution of the compound they have formed. This latter solution does not, like the hydrochloric acid solution, redden litmus-tinted paper; nor does it, like the ammonia solution, restore the blue colour to litmus-paper reddened by acidulation. With two bodies of strongly marked, and decidedly opposite characters, we have formed a neutral compound, in aspect and general characters strongly resembling common culinary salt. This product, in conformity with a very usual artifice of chemical nomenclature, is distinguished by a name derived from the appellations of its two constituents; it is called, at pleasure, either *ammonia-hydrochlorate*, or *hydrochlorate of ammonia*.

This salt-like body is readily volatilizable, and were it not subject to the phenomenon of dissociation, to which our attention has already been directed, we should have no difficulty in establishing its normal molecular constitution as follows:—

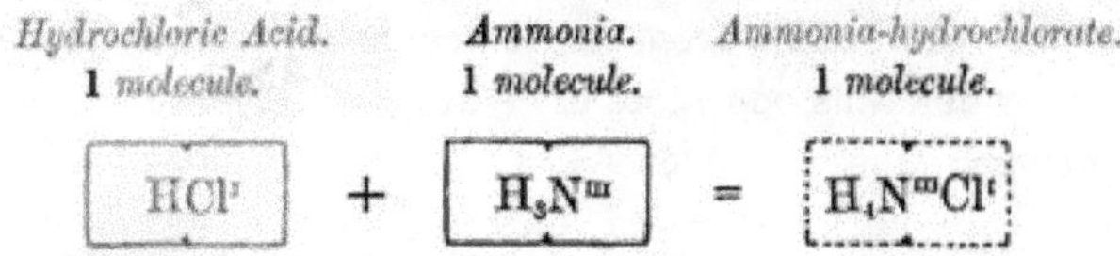

The dimensions of the molecule are here marked in dotted outline, our sign of doubt. But of the ponderal constitution of this compound we are in no doubt at all. We know that this product is a ternary compound, consisting of hydrogen, 4 vols., united with nitrogen and chlorine, one volume each. If, as we assume, these 6 volumes of constituent gases condense into the normal dilitral product-volume, we have here a condensation ratio of $\frac{2}{6} = \frac{1}{3}$.

But, while we provisionally adopt this view, we must carefully bear in mind that it rests upon analogy only, not on fact. Every attempt hitherto made to obtain the assumed normal vapour, by volatilizing hydrochlorate of ammonia, has resulted in the production of a quantity of gas equal to twice the normal product-volume. In other words, the volume obtained is equal to the sum of the volumes of the two constituent gases, in their free state. These are accordingly believed to dissociate during the act of vaporization of the ammonia-hydrochlorate.

But our early experiments supply us with other examples of ternary compounds, which call in their turn for brief consideration here.

In studying the composition of hydrochloric acid, water and ammonia, we employed, as you remember, an alkali-metal, preferably sodium, to liberate the hydrogen from those several compounds; the final result being, in each case, that the sodium replaced the hydrogen set free, and combined in its stead with the chlorine of the hydrochloric acid, with the oxygen of the water, and with the nitrogen of the ammonia.

Now, as chlorine, oxygen, and nitrogen are respectively uni-, bi-, and tri-valent; the molecules of their hydrogen compounds contain respectively (as we well remember) one, two, and three atoms of hydrogen.

Evidently, therefore, from the molecule of hydrochloric acid, which only contains one atom of hydrogen, either *all* or *none* of the hydrogen must be displaced.

From the molecule of water, on the contrary, containing as it does two atoms of hydrogen, it is possible to conceive the displacement of those atoms by two successive invasions (so to speak) of the alkali metal; one atom, or half the whole quantity, being first displaced, and afterwards the remainder.

Again, from the molecule of ammonia, with its three atoms of hydrogen, we readily understand that sodium may successively displace first one, and then a second, of those hydrogen atoms, before finally displacing the third.

Now, though, at the time of making these experiments, we were only interested in, and therefore only dwelt on, the final result, viz., the complete expulsion of the hydrogen from these three compounds, and the consequent formation of binary compounds only between the sodium and the chlorine, the oxygen, and the nitrogen respectively; it now becomes opportune to dwell on the fact that, in the cases of water and ammonia, the final result is not attained at once, but that the above-mentioned intermediate ternary compounds are in reality produced during the successive stages of the process.

The ternary products of the progressive dehydrogenation of water and ammonia by sodium are, in the diagram subjoined, placed intermediately, between the original compounds, water and ammonia, which are figured on the left, and the final products, oxide and nitride of sodium, which appear to the right. For clearness' sake the atomic constitution of each product is displayed without condensation.

Progressive Decomposition of Water and Ammonia.

ORIGINAL. (*Binary*).	INTERMEDIATE. (*Ternary*).		FINAL. (*Binary*).
H, O^{II}, H		Na^{I}, O^{II}, H	Na^{I}, O^{II}, Na^{I}
		(*hypothetical*)	
H	Na^{I}	Na^{I}	Na^{I}
H N^{III}	H N^{III}	Na^{I} N^{III}	Na^{I} N^{III}
H	H	H	Na^{I}

The one ternary compound here shown between water and soda consists, as you observe, of one atom each of sodium, hydrogen, and oxygen. It may be considered as water, in which half the hydrogen is replaced by sodium. It is a substance well known to us all, being the common hydrated or caustic soda, so commonly employed in the arts and manufactures. The final binary product, from which all the hydrogen has disappeared, sodium alone remaining in combination with oxygen, only retains, of the parent-compound water, from which it sprung, the characteristic structural type. It is, therefore, often called sodic anhydride, meaning literally *water-less soda*; for which the term dry oxide of sodium is often substituted. These remarks, as well as the diagrammatic representations themselves, are as applicable to potassium as to sodium, and may be converted into potassic expressions by merely substituting K for Na throughout. Or the initial M (for metal) may be used to give the expressions a general form, coinclusive of both the alkali-metals—an artifice of symbolic generalization very commonly employed by chemists.

In the ammonia-sodic series, figured below the aqua-sodic series in the diagram, we see two ternaries, one unisodic, the other bisodic, intervening between the unmodified ammonia and the nitride of sodium which is the final product of its transformation. The bisodic compound is hypothetical.

The reactions by which, in the case of water and ammonia, these serial products are obtained, each in succession by the enrichment of a previous one with additional sodium, are displayed in the following molecular equations, which also show the proportion of hydrogen liberated at each stage of the operation :—

1. *Progressive Inception of Sodium by Water.*

1st Stage .. $2\,[H_2O^{II}] + [Na^INa^I] = 2\,[Na^IHO^{II}] + [HH]$

2nd Stage .. $2\,[Na^IHO^{II}] + [Na^INa^I] = 2\,[Na^I_2O^{II}] + [HH]$

2. *Progressive Inception of Sodium by Ammonia.*

1st Stage .. $2\,[H_3N^{III}] + [Na^INa^I] = 2\,[Na^IH_2N^{III}] + [HH]$

2nd Stage .. $2\,[Na^IH^2N^{III}] + [Na^INa^I] = 2\,[Na^I_2HN^{III}] + [HH]$

3rd Stage .. $2\,[Na^I_2HN^{III}] + [Na^INa^I] = 2\,[Na^I_3N^{III}] + [HH]$

These equations bring pointedly under our notice the remarkable fact, that, in the progressive transformation of ammonia by sodium, each additional atom of metal that enters into the compound *takes the place* of a hydrogen-atom; which is, as it were, expelled to make room for it.

Our fourth typical group, that headed by marsh-gas, affords us one case, and one only, of a ternary compound exemplifying this peculiar mode of formation; that is to say, resulting from the displacement of a hydrogen-atom by a sodium-atom entering the molecule in its stead. The formula of this ternary compound is—

$$Na^{I}H_3C^{IV}$$

This product is formed with considerable difficulty, and by circuitous processes, upon which we cannot here dwell. We refer to it only for the sake of remarking that each of our four typical groups exemplifies the formation of ternary compounds by the expulsion of hydrogen, and the replacement of the atom or atoms so expelled by a quantivalent proportion of sodium.

By these particular examples we are led up to the general and most important conception of *Substitution-compounds;* that is to say, of bodies formed (often in extensive series) by the replacement of one or more of the constituent atoms of a compound by atoms of some other body introduced in their stead. And we thus make acquaintance, in germ, with a principle, from which, as from a living seed, the mighty growth of modern chemistry has mainly sprung.

This principle is too important to be passed over lightly: it deserves the best elucidation that the experiments which we have already tried together enable us to bestow on it.

Among our early experiments on water, ammonia, and marsh-gas, we remember effecting their decomposition by employing chlorine to withdraw from them their hydrogen atoms, so as to liberate from water oxygen, from ammonia nitrogen, and from marsh-gas carbon. (Comp. pp. 28, 33, and 102.)

At that stage of our inquiry it would have served no useful purpose to point out that, under certain conditions of procedure, the withdrawal of hydrogen from those compounds by chlorine takes place in the same progressive manner as we have seen it affect when sodium was the displacing agent employed. This fact now interests us deeply as a corroboration of the great law of substitution, which is dawning on our view.

The reaction in question takes place in the following manner. While part of the chlorine is withdrawing hydrogen from water, ammonia, or marsh-gas, as the case may be, and forming hydrochloric acid with that displaced hydrogen, another portion of chlorine takes the place of the said hydrogen in the compound from which it is expelled, the final products of the transformation being—

In the case of water*—

$$Cl^{\text{i}}_2O^{\text{ii}}, \text{ instead of } H_2O^{\text{ii}}.$$

In the case of ammonia—

$$Cl^{\text{i}}_3N^{\text{iii}}, \text{ instead of } H_3N^{\text{iii}}.$$

In the case of marsh-gas—

$$Cl^{\text{i}}_4C^{\text{iv}}, \text{ instead of } H_4C^{\text{iv}}.$$

Evidently, here also, as in the decomposition by sodium, we may expect the formation of intermediate ternary compounds, due to the partial and progressive replacement of the hydrogen atoms by atoms of the attacking agent.

In the following diagrams we have (as before) placed between the original and final binary compounds the ternary compounds capable of intermediate production.

---

* Practically, the metallic derivatives of water are employed for this purpose; their decomposition by chlorine being more readily effected than that of water itself.

Compounds.

ORIGINAL. (*Binary*).	INTERMEDIATE. (*Ternary*).			FINAL. (*Binary*.)
H, O^{ii}, H	Cl^{i}, O^{ii}, H			Cl^{i}, O^{ii}, Cl^{i}
H, H N^{iii}, H	Cl^{i}, H N^{iii}, H	Cl^{i}, Cl^{i} N^{iii}, H		Cl^{i}, Cl^{i} N^{iii}, Cl^{i}
H, H, C^{iv}, H, H	Cl^{i}, H, C^{iv}, H, H	Cl^{i}, Cl^{i}, C^{iv}, H, H	Cl^{i}, Cl^{i}, C^{iv}, Cl^{i}, H	Cl^{i}, Cl^{i}, C^{iv}, Cl^{i}, Cl

From this diagram we gather, at a glance, that between water and the bichloride of oxygen there intervenes but one ternary compound; while, between ammonia and terchloride of nitrogen, two ternaries find place, and between marsh-gas and tetrachloride of carbon, no less than three ternaries are interposed. The intermediate ammoniacal compounds have never yet, it must be admitted (partly on account of their dangerously explosive character), been produced in sufficient purity and abundance for analysis; though, of their existence, analogy scarcely permits a doubt. All the other compounds symbolized are bodies of well-ascertained composition and properties.

The production of these substitution compounds, intermediate and final, each from its predecessor, by additional inception of the invading element, at the expense of hydrogen expelled and converted into hydrochloric acid, may be shown as before by ordinary equations; which fall, in this case, into three sets, respectively representing the progressive transformation of water, ammonia, and marsh-gas.

TERNARY CHLORINE-DERIVATIVES:

1. *Of Water.*

$$H_2O + ClCl = HCl + HClO$$

$$HClO + ClCl = HCl + Cl_2O$$

2. *Of Ammonia.*

$$H_3N + ClCl = HCl + H_2ClN$$

$$H_2ClN + ClCl = HCl + HCl_2N$$

$$HCl_2N + ClCl = HCl + Cl_3N$$

3. *Of Marsh-gas.*

$$H_4C + ClCl = HCl + H_3ClC$$

$$H_3ClC + ClCl = HCl + H_2Cl_2C$$

$$H_2Cl_2C + ClCl = HCl + HCl_3C$$

$$HCl_3C + ClCl = HCl + Cl_4C$$

We have not yet, however, learnt from these diagrams all that they can teach us. They shadow forth clearly another and a most general fact of modern chemistry, viz., the uniform retention, by substitutional derivative compounds, of the structural type affected by their primary or parent compound.

We must not dwell here on this pregnant theme—the law of atomic substitution; nor, indeed, have we acquired facts enough, as yet, to illustrate its nature, and prove its importance. A moment's reflection will, however, enable us to conceive how vast must be the aid derivable from such a principle in the fulfilment of the task, else almost impracticable, of distributing into a system of natural groups the ever-growing multitudes of serial compounds.

From the foregoing remarks and examples, we have learned that a binary compound molecule may become ternary, either by uniting with another *molecule* (as in the case of the ammonia-hydrochlorate), or by receiving into its structure one or more *atoms*, of a third body in replacement of a proportion of its own atoms. In this latter class of cases it is found that the number of quantivalents brought in by the atoms introduced is always exactly equal to the number of quantivalents carried away in the atoms expelled. It is true that, in the examples given, we have only employed univalent chlorine and sodium to replace, atom for atom, univalent hydrogen. We shall, however, hereafter learn that this law of substitutional quantivalence is universal, and holds good in all cases of substitutional reaction, whether the incoming and outgoing atoms be uni-, bi-, tri-, or quadrivalent, or be partly of one class and partly of another. The law of which we thus obtain a first glimpse is one of the most important results of modern chemical research, and it will be continually pressed on our notice during the course of our future studies.

There is yet a third manner, besides the two just mentioned, in which a binary compound molecule may become ternary; and this consists in its direct assumption of one or more atoms of a third element, without loss, by substitutional displacement, of any atom or atoms of its own.

Oxygen, for example, is thus taken up by the majority of our typical binary hydrogen-compounds and their congeners. Hydrochloric acid, the type of our first group, may be cited as possessing this faculty in a high degree. Its molecule can take up 1, 2, 3, or 4 atoms of oxygen, and form therewith a series of four well-marked ternary compounds, which may be called the hydrochloroxygen series, and which are represented by the following expressions :—

Hydrochloroxygen Series.

$$HCl'O'', \quad HCl'O''_2, \quad HCl'O''_3, \quad HCl'O''_4.$$

The congeners of hydrochloric acid, hydrobromic and hydriodic acids possess similar properties, and both of them form series of oxygen compounds, analogous to the hydrochloroxygen series above.

This faculty of taking up oxygen is by no means wanting in our second group of typical compounds, though it is more conspicuous in the congeners than in the architype, water itself.

Water, indeed, does form one compound, by the inception of additional oxygen; but the product is not easy to obtain, nor is it of stable character. But the analogues of water—the hydrogen compounds of sulphur and selenium—form, each of them, two well-known ternary compounds, by incorporating, in their respective molecules 3 atoms and 4 atoms of oxygen; whence we may reasonably presume that analogous intermediate combinations of these bodies with 1 and 2 atoms of oxygen are possible, and may one day be obtained. The entire series, actual and possible, of the sulphuretted-hydrogen oxides is displayed in the subjoined set of expressions :—

Hydrosulphoxygen series.

$$H_2S''O'', \quad H_2S''O''_2, \quad H_2S''O''_3, \quad H_2S''O''_4.$$

The two first formulæ of this series represent its as yet hypothetical members. The last two, on the contrary, represent, with perfect accuracy, the composition of two of the most important acids known to chemistry, as well scientific as industrial, viz., the monohydrated sulphurous and sulphuric acids. These also we must pass in our rapid course; reserving for a more fitting occasion the cultivation of their valuable acquaintance.

Passing to our third typical group, headed, as we remember, by ammonia, and comprising among its members phosporetted and arsenetted hydrogen, we find that, though ammonia itself has not yet been induced to combine with oxygen, the analogue of ammonia, phosphoretted hydrogen, is conspicuous for the number and variety of its ternary compounds with that body; combinations of the molecule $H_3P^{'''}$ with 2, 3, and 4 atoms of oxygen having actually been obtained, so that only its possible protoxide remains to be produced to complete the quadruple series. The formation and composition of these ternaries, including the as yet hypothetical member of the series, are displayed in the following expressions:—

Hydrophosphoxygen series.

$$H_3P^{'''}O^{''}, \quad H_3P^{'''}O^{''}_2, \quad H_3P^{'''}O^{''}_3, \quad H_3P^{'''}O^{''}_4.$$

Our fourth and last typical group, at the head of which stands marsh-gas, has hitherto furnished only one example of a ternary oxygen compound; which has the formula $H_4C^{''''}O^{''}$. This compound, however, which we shall hereafter study under the name of Methylic Alcohol, makes up, by its extreme importance, for the else sterile character of the marsh-gas group, in respect of ternary oxygen-derivatives. Through methylic alcohol, as through a widely opened gate, we shall pass to the study of its almost innumerable derivatives, which form in themselves one of the largest tracts of that domain through which our future journey lies.

But now, fellow-travellers, we have arrived at a point where our progress together must be interrupted for a time. The plan of this brief introductory course is fulfilled; and its objects so far as my limited powers permit their achievement, are attained. Before we part, however, let us, as wayfarers are wont, rest a little while on the gentle eminence we have attained, and take a retrospective survey of our route thus far. In other words—and to drop metaphor—let us endeavour, by a summary recapitulation, to fix in our memory some of the leading facts and principles which our experiments and reasonings, our inductions and deductions, have gradually unfolded to our view.

As our point of departure we selected, you remember, the familiar fluid, *water;* of which we learned the compound nature by an experiment, the simplest, perhaps, and the most striking, in the whole range of chemistry.

This consisted in dropping on its surface a fragment of the alkali-metal, potassium; which, at the touch of water, took fire. at the same time liberating from it an inflammable gas called hydrogen. This, upon examination, proved to be the lightest body known; and we accordingly adopted it as our unitary standard of volume-weight.

This standard gas we proceeded to liberate, by the same simple means, from two other bodies, less familiar than water, but very well known in the arts and manufactures, viz., muriatic acid and ammonia, both, in their pure state, gases at ordinary temperature and pressure.

The further examination of these three sources of hydrogen revealed to us the existence of three gaseous bodies, chlorine, oxygen, and nitrogen, as being respectively associated with hydrogen, in muriatic acid, water, and ammonia.

The study of chlorine, oxygen, and nitrogen, showed us, in the first, one of the most active of chemical agents; in the second, the typical supporter of combustion; in the third, one of the most inert bodies known.

By the *analytic* method, applied to decompose the three hydrogen compounds of these gases, and by the *synthetic* method, employed, so far as available, to reconstruct them, we learned the proportions, as well by volume as by weight, in which chlorine, oxygen, and nitrogen respectively combine with hydrogen, in muriatic acid, water-gas, and ammonia.

With one unit-volume each of chlorine, oxygen, and nitrogen, weighing respectively 35·5, 16, and 14, we found hydrogen combining in the unit-volume and weight ratios of 1 for the first-named body, 2 for the second, and 3 for the third.

Notwithstanding this inequality in the number of unit-volumes of the gaseous constituents of these compounds, we found the volumes of the gaseous products to be exactly equal; measuring, in all three cases alike, 2 unit-volumes. This curious circumstance proved to us that condensation increases, in these typical cases, *pari passu* with the number of hydrogen-volumes engaged.

We thus experimentally established three well-defined models of chemical structure, displayed in the combination of 3 typical elements with the standard element, hydrogen.

To these types, both of elementary and compound bodies, a fourth in each kind was soon afterwards added; *carbon* presenting itself to our notice as the type of the *non-volatile* elements; while the richest in hydrogen of its hydrogen-compounds, viz., *marsh-gas*, contributed its final term to our series of structural models. We had found the three gaseous typical elements successively engaging, within equal product-volumes (double the unit-volume in each case), 1, 2, and 3 volumes of hydrogen; and now, in the like product-volume of marsh-gas, we found carbon engaging 4 volumes of hydrogen.

Hence the sort of *disjunctive conjunction* by which we annexed to our typical series of compounds this singular body, marsh-gas; alien, with regard to the non-volatile character of its typical constituent, carbon; cognate, in respect of its ponderal, volumetric, and condensational relations with its standard constituent, hydrogen.

To the establishment of the typical elements, and their typical hydrogen-compounds, succeeded naturally the study of the congeners in each kind; and we were thus led to make acquaintance with bromine and iodine as analogues of chlorine, and with their respective hydrogen-compounds as analogues of hydrochloric acid. With oxygen and its hydrogen-compound, water, we associated, in like manner, sulphur and selenium, and their respective hydrogen-compounds, cast in the structural mould of water-gas. With nitrogen and its hydrogen-compound, ammonia, we connected phosphorus and arsenic, and their ammonia-like combinations with hydrogen. With carbon, lastly, and its hydrogen-compound, marsh-gas, we conjoined, in like manner, silicon, and its hydrogen-compound, formed on the marsh-gas model.

In these four groups of typical elements and compounds, we recognized the germ of a grand conception—that of a natural classification of chemical bodies into genera and species, each distinguished by well-marked characteristics, not excluding individual varieties, but grouping them in subordination to collective laws.

In the course of these experimental demonstrations, we became acquainted with the meaning of the term *chemistry*, and we obtained our first notions concerning the nature of chemical phenomena. We learned, for example, the characters of elementary as contradistinguished from compound bodies; of chemical combination as contradistinguished from mere mechanical mixture; of combining proportions, volumetric and ponderal, and of the immutability by which they are characterized.

While thus gradually learning the general principles and laws of chemistry, we also became familiarized, as we proceeded, with the aspect and uses of chemical apparatus, and with the manipulations necessary for their dexterous employment. We gained experience more particularly of the methods in use for generating, collecting, transferring, measuring, desiccating, testing, and weighing gaseous bodies; and for ascertaining the influence of varying temperature and pressure on their bulk and density.

Our analytic and synthetic operations obliged us to employ, by turns, the powerful agencies of electricity, light, heat, and the specific power we termed *chemism*, in order to bring about the desired reactions; and these we often found to be attended with remarkable physical perturbations; as, for instance, with the sudden development of light and heat, and often with more or less violent explosion. The means of generating and applying the (so-called) imponderable forces, and of controlling their effects, when excessive or dangerous, were thus brought prominently, though cursorily, under our notice.

Upon the individual characters of the elements and compounds thus submitted to investigation we did not dwell at length; nor, indeed, did we enlarge even upon the general laws with which we met in our course: it was our care to note only such particulars as came within the scope, and promoted the purpose, of our immediate inquiry.

Doubtless, each subject which we thus touched by the way opened a tempting path to our curiosity; as the climber, whose appointed aim is the top of the tree, is tempted by the fruit-laden branches he passes in his ascent.

But though we gladly accepted the incidental information which our experiments naturally threw in our way, we forbore from prolonged digressions, and persisted steadily in the straight course of our inquiry.

This led us next to the study of the curious and important bodies constituting the nitroxygen series — a study in which we broke entirely new ground; quitting the consideration of the typical hydrogen-compounds, each of which only exemplifies combination in a single fixed ratio, and advancing to the examination of a new and pregnant law of chemistry,—that of combination in *multiple proportions.*

As our induction thus extended itself, and our facts began to accumulate, we felt the want of some instrument of record, less periphrastic than ordinary parlance, to epitomise concisely, and to bring graphically and simultaneously under the eye, trains of phenomena which it would else be difficult to grasp and comprehend, firstly, in their mutual relations to each other, and, secondly, in their common dependence on general laws.

We were thus induced to represent our gas-volumes by squares, *lettered* with the initials of the bodies depicted, *figured* with their relative volume-weights, and forming the germ of a symbolic nomenclature and notation which, in the further stages of our progress, we were enabled at once to enrich and to simplify, by incorporating in it our newly-acquired facts, and eliminating from it forms too cumbersome for practical use, though invaluable as aids to chemical education.

That our proportional numbers, abstract at first, might acquire a concrete significance, we had to make choice of some system of weight and measure, in terms of whose unitary standards to express those otherwise vague determinations.

This led us to study the admirable metrical system of the French, which supplied us with our standard of capacity, the *litre*, and of weight, the *gramme*; at the same time teaching us, by means of Greek and Latin prefixes, to express their multiplication and division in decimal progression.

The weight of the standard volume (1 litre) of our standard element (hydrogen), expressed in terms of our new ponderal unit (1 gramme), gave us the invaluable coefficient 0·0896 gramme, which, as you remember, we called our "barley-corn-weight" or *crith*, by means whereof, as a multiplier, we convert the figures representing the mere abstract specific gravities of the various gases and vapours into expressions of their actual or concrete litre-weights.

The concrete values with which our symbolic expressions thus became clothed increased the power of our system of notation, both as a language for recording phenomena, and as an instrument to assist in their investigation, experimental and theoretic.

Thus armed, we ventured upon speculative ground; we sought the interpretation of the phenomena we had as yet not observed; we endeavoured to explain, by a rational hypothesis, the remarkable proportionality, ponderal and volumetric, of chemical reactions; and, with this view, we entered on the inquiry, What *is* matter? Of what parts is it composed? How are these affected by the solid, fluid, and gaseous conditions?

How are their interspaces filled? And what, in particular, is the corpuscular construction of a gas?

In studying these questions, we were led to admit the three-fold divisibility, molar, molecular, and atomic, of material bodies, and to refrain from asserting their *infinite* divisibility. The elasticity of gases we attributed to a force connected, in some unknown way, with *heat*; whose specific relations to different bodies, and so-called *latency* therein, enabled us to understand certain, else inexplicable, properties of gases, and to conceive these bodies as built up of molecules, or atom-clusters, of which all gases are assumed to contain equal numbers in equal volumes.

In the light of these conceptions our symbolic language took on a new significance. Besides representing volumes and volume-weights, our squares became pictures for us of molecules and atoms; whose movements of decomposition and reconstruction we were thus, in imagination, enabled to follow.

We thus became familiar with the diatomic structure of the typical elementary molecules; with the monatomic and poly-atomic structure of certain other molecules; and with the molecule-forming and atom-fixing powers of the elementary atoms; the former of which powers we found to be in the ratio of their atom-weights, while the latter we identified with their atom-freeing and atom-replacing powers, all of which we included in the term quantivalence. These studies led us naturally to touch on and illustrate the principles of quantivalential equipoise, and to embody the idea of quantivalence in our symbolic language, by the addition thereto of appropriate quantivalential coefficients.

Having learned this much from the study of binary compounds, we passed on to consider compounds of a higher order, ternary, quaternary, &c., and the several modes of their genesis from binary compounds; as, for instance, sometimes by molecular, sometimes by atomic inception, each sometimes attended, and sometimes not, by substitutional displacement of atoms from the parent compound.

Examples of ternary compounds generated in each of these modes were supplied to us, in the lecture of to-day, by the four typical groups with whose study we commenced our course. In our rapid review of these compounds, we noted their usually closely-packed product-volumes, or high vapour-densities; their frequent tendency to dissociation; their habitual retention of the structural type of their parent compounds; and the principles of their progressive or serial development. The last of these interesting compounds was presented to us by marsh-gas; which, in becoming methylic alcohol by inception of oxygen, threw open the gate of a new field of inquiry, and brought our present journey to its term.

In this rapid retrospective survey, I have not attempted to be encyclopædic; my wish has merely been, at parting, to recal the more important of the many deeply-interesting topics which flowed upon us in succession, as out of a living source, from the pregnant conception of the molecular and atomic construction of matter.

It can now, indeed, be no matter of surprise to you that we devoted so much time to the consideration of molecules and atoms, and dwelt at so much length upon the methods in use for determining the molecular and atomic weights of the elements. These are the foundations of chemical knowledge; and the table in which they are given, together with their symbols and coefficients of quantivalence, though, as we are aware, some of the figures are still doubtful, deserves our close and frequent study. The more of its figures we can bear in mind, the more accurate and ready will our knowledge be, whether for theoretic or practical applications.

But I must not linger on these themes; nor, to defer unwelcome separation, trespass still further beyond limits already overpast.

If, in conclusion, I resume my metaphor, and bid you adieu as fellow-travellers, it is because I deeply feel how much there is in the present transitional condition of chemistry to justify such an expression; and am almost painfully conscious how narrow is the explored domain through which the teacher can be your guide, in comparison with those vast regions of truth as yet unknown, in which we are all fellow-students together!

And thus, a learner myself, day by day, I can the better appreciate your constant and sympathetic attention to my lessons; and am encouraged the more earnestly to hope that the facts and principles, which, in these few meetings, we have passed in review together, will not merely afford us some present insight into the new doctrines now so deeply, and, let me add, so wholesomely, agitating the chemical world, but will also serve as a firm basis on which, in future conferences, we may build up, stone by stone, the vast superstructure of the chemical edifice.

THE END.

LONDON: PRINTED BY WILLIAM CLOWES AND SONS, STAMFORD STREET
AND CHARING CROSS.

www.ingramcontent.com/pod-product-compliance
Lightning Source LLC
Chambersburg PA
CBHW021948220326
41599CB00012BA/1381

* 9 7 8 3 3 3 7 0 3 8 9 4 6 *